Wireless Networks

Series Editor

Xuemin Sherman Shen, University of Waterloo, Waterloo, ON, Canada

The purpose of Springer's Wireless Networks book series is to establish the state of the art and set the course for future research and development in wireless communication networks. The scope of this series includes not only all aspects of wireless networks (including cellular networks, WiFi, sensor networks, and vehicular networks), but related areas such as cloud computing and big data. The series serves as a central source of references for wireless networks research and development. It aims to publish thorough and cohesive overviews on specific topics in wireless networks, as well as works that are larger in scope than survey articles and that contain more detailed background information. The series also provides coverage of advanced and timely topics worthy of monographs, contributed volumes, textbooks and handbooks.

** Indexing: Wireless Networks is indexed in EBSCO databases and DPLB **

More information about this series at http://www.springer.com/series/14180

Reginald A. Banez • Lixin Li
Chungang Yang • Zhu Han

Mean Field Game and its Applications in Wireless Networks

Reginald A. Banez
Department of Electrical and Computer
Engineering
University of Houston
Houston, TX, USA

Chungang Yang
School of Telecommunications Engineering
Xidian University
Xi'an, Shaanxi, China

Lixin Li
School of Electronics and Information
Northwestern Polytechnical University
Xi'an, Shaanxi, China

Zhu Han
Department of Electrical and Computer
Engineering
University of Houston
Houston, TX, USA

ISSN 2366-1186   ISSN 2366-1445   (electronic)
Wireless Networks
ISBN 978-3-030-86907-6   ISBN 978-3-030-86905-2   (eBook)
https://doi.org/10.1007/978-3-030-86905-2

This Springer imprint is published by the registered company Springer Nature Switzerland AG
The registered company address is: Gewerbestrasse 11, 6330 Cham, Switzerland

Preface

The current generation of wireless networks is approaching its limits caused by increasing data traffic, more frequent network usage, and rising number of connected devices. In order to overcome these limitations, enabling technologies such as ultra-dense networks, multi-access edge networks, and massive antenna arrays are proposed as part of the future generation of wireless networks. However, in order to analyze, model, and simulate these technologies, an appropriate mathematical framework that can handle a large number of interacting entities is necessary. Hence, this book focuses on mean field games (MFGs) and their applications in future wireless networks.

MFGs deal with the study and analysis of differential games with infinitely many players. The theory of MFG enables the study of the Nash equilibrium of games with very large number of indistinguishable players. It allows a player to make a decision or strategy based on the state distribution of all the players instead on the individual states of other players. Meanwhile, a mean-field-type game (MFTG), a subclass and relaxed version of MFG, has been applied to applications where the MFG assumptions do not necessarily hold. In MFTG, the number of decision makers may be infinite or finite, the decision makers may not be indistinguishable, and a decision maker may have a significant effect on the state distribution.

This book starts with an overview of the current and future state-of-the-art in 5G and 6G wireless networks. Next, a tutorial on MFG, MFTG, and prerequisite fields of study, such as optimal control theory and differential games, is presented. Afterwards, several applications of MFG and MFTG in ultra-dense networks, social networks, and multi-access edge computing networks are introduced.

Furthermore, the goal of this book is to educate electrical and computer engineers as well as applied mathematicians about the significance of MFG and MFTG in analyzing and designing future wireless networks.

Houston, TX, USA
Xi'an, China
Xi'an, China
Houston, TX, USA

Reginald A. Banez
Lixin Li
Chungang Yang
Zhu Han

Contents

Acronyms

2D	Two Dimensional
3D	Three Dimensional
4G	Fourth Generation
5G	Fifth Generation
A2A	Air-to-Air
A2G	Air-to-Ground
AEA	Average Estimation Accuracy
AF	Amplify-and-Forward
AP	Access Point
AQI	Air Quality Index
ATC	Air Traffic Control
AWGN	Additive White Gaussian Noise
BS	Base Station
CDF	Cumulative Distribution Function
CNPC	Control and Non-Payload Communication
CU	Cellular User
D2D	Device-to-Device
DC	Difference of Convex Functions
DF	Decode-and-Forward
DQN	Deep Q-Network
EE	Energy Efficiency
eMBB	enhanced Mobile Broadband
F-Cell	Flying-Cell
GPI	Generalized Policy Iteration
GPM-NN	Gaussian Plume Model Embedding Neural Networks
GSC	Ground Control Station
IC	Incentive Compatibility
IoT	Internet-of-Things
IP	Increasing Preference
IR	Individual Rationality

ITU-R	International Telecommunication Union–Radiocommunications Standardization Sector
KKT	Karush-Kuhn-Tucker
LoS	Line-of-Sight
LTE	Long-Term Evolution
MBS	Macro-cell Base Station
MD	Mobile Device
MDP	Markov Decision Processes
mMTC	massive Machine-Type Communications
NLoS	Non-Line-of-Sight
NN	Neural Networks
PC	Payload Communication
PDF	Probability Density Function
PDT	Partial Derivative Threshold
PoI	Point of Interests
PPP	Poisson Point Process
QoS	Quality of Services
RMa	Rural Macro
SBS	Small-cell Base Station
SINR	Signal-to-Interference-plus-Noise Ratio
SNR	Signal-to-Noise Ratio
SVM	Support Vector Machine
TD	Temporal-Difference
TSP	Traveling Salesman Problem
U2N	UAV-to-Network
U2U	UAV-to-UAV
U2X	UAV-to-Everything
UAS	Unmanned Aircraft Systems
UAV	Unmanned Aerial Vehicle
UE	User Equipment
UMa	Urban Macro
UMi	Urban Micro
URLLC	Ultra-Reliable and Low-Latency Communications

Chapter 1
Overview of Mean Field Games in Wireless Networks

The current generation of wireless networks allows users to access a variety of services such as data, voice, and multimedia. However, according to the latest Cisco Annual Internet Report, these networks are reaching their full potential caused by growth of data traffic, more frequent network usage, and rise in number of devices and connections. Consequently, these circumstances have inspired interest and necessity for a new generation of wireless networks that can support the growing number of devices and connections as well as the volume of data they generate.

The main goal of the next generation of wireless networks is to improve the implementation of current wireless networks as well as the services provided to the network users. In 5G, the networks are required to handle high capacity, high data rate, low latency, massive connectivity, increased network reliability, and improved energy efficiency. In 6G, the networks are required to improve over the 5G networks in order to support new applications such as multisensory extended reality (XR), wireless brain-computer interactions (BCI), connected robotics and autonomous systems (CRAS), and blockchain and distributed ledger technologies (DLT).

Several technologies have been proposed to meet the requirements of the next generation of wireless networks. For 5G, some of these technologies include massive MIMO, millimeter wave communications, ultra-dense networks, spectrum sharing, device-to-device communications, full-duplex communications, and cloud technologies. For 6G, the proposed technologies are communication with large intelligent surfaces, integrated terrestrial, aerial, and satellite networks, and energy transfer and harvesting.

In this chapter, an overview of the 5G/6G wireless networks is discussed including the motivation for such network upgrades and the technical requirements of these networks in order to deliver satisfactory services to end users. Furthermore, wireless technologies that can accomplish delivery of 5G/6G quality services to end users are also considered. Afterwards, mean field games (MFGs) are introduced. These are mathematical frameworks that can simplify the modeling and analysis of

© The Editor(s) (if applicable) and The Author(s), under exclusive license to Springer Nature Switzerland AG 2021
R. A. Banez et al., *Mean Field Game and its Applications in Wireless Networks*, Wireless Networks, https://doi.org/10.1007/978-3-030-86905-2_1

complicated 5G/6G network structures. Finally, an introduction of several research works on mean field game applications in wireless networks is presented.

1.1   Background and Requirements

According to the latest Cisco Annual Internet Report [1], the current wireless network infrastructures are reaching their full potential in terms of data rate, bandwidth, and capacity caused by growth of data traffic, more frequent network usage, and rise in number of devices and connections. Consequently, these circumstances have inspired interest and necessity for a new generation of wireless networks that can support the growing number of devices and connections as well as the volume of data they generate.

The next generation of wireless networks is envisioned to upgrade the current implementation of wireless networks. 5G networks are expected to handle high capacity, high data rate, low latency, massive connectivity, increased network reliability, and improved energy efficiency [2]. These networks are envisioned to support 5G applications such as mobile computing, advanced vehicular communications, remote health monitoring, smart home, smart cities, and smart grid. Meanwhile, 6G networks are expected to improve over the 5G networks and support new applications such as multisensory extended reality (XR), wireless brain-computer interactions (BCI), connected robotics and autonomous systems (CRAS), and blockchain and distributed ledger technologies (DLT) [3].

In order to fulfill these network requirements, enabling technologies have been proposed. For 5G, these technologies include massive MIMO, millimeter wave communications, ultra-dense networks, spectrum sharing, device-to-device communications, full-duplex communications, and cloud technologies [4]. For 6G, the proposed technologies are communication with large intelligent surfaces, integrated terrestrial, aerial, and satellite networks, and energy transfer and harvesting [3].

1.1.1   Technical Requirements

The authors of [5] compiled the major requirements of next generation 5G systems from the works in [6, 7], and [8]. In 5G networks, the data rates are expected to be within 1–10 Gbps, which is 10 times the 150 Mbps theoretical data rate of a 4G LTE network. The round trip latency is 1 ms, which is near 10 times smaller than the 10 ms round trip latency in 4G. The 5G networks must be able to provide connectivity to an enormous number of connected devices. Consequently, the 5G network bandwidth must be high in order to service these connected devices. The 5G network must have an availability of 99.999% and a coverage of almost 100%. The energy usage is suggested to be reduced by almost 90%. Thus, for devices using battery, the battery life should be high.

Table 1.1 Summary of technical requirements for 5G and 6G wireless networks

Technical requirements	5G	6G
Peak data rate	20 Gbps	1 Tbps
End-to-end delay	1 ms	less than 1 ms
Processing delay	100 ns	10 ns
Reliability	99.999%	99.99999%
Area traffic capacity	10 Mbps/m^2	1 Gbps/m^2
Frequency bands	Sub-6 GHz mmWave (24–52.6 GHz)	Sub-6 GHz mmWave THz band
Connection density	One million devices/km^2	Ten million devices/km^2
Mobility	500 km/h	greater than 700 km/h

According to IMT 2020, the technical requirements for 5G technologies are as follows [9]. The peak download data rate is 20 Gbps, and the peak uplink data rate is 10 Gbps. The peak download spectral efficiency is 30 bps/Hz, while the peak uplink spectral efficiency is 15 bps/Hz. The user experienced data rates are 100 and 50 Mbps in the downlink and uplink, respectively. The bandwidth is at least 100 MHz and up to 1 GHz for operations in higher frequency bands.

6G networks are expected to gain more attention starting in 2030. However, research for 6G networks has already begun in some countries. In order to support 6G applications, 6G networks must be able to deliver around 1 Tbps, which is 1000 times the data rate in 5G networks. Moreover, the round trip latency must be less than 1 ms to provide satisfactory service to end users. The reliability requirement in 6G networks is 99.99999%. In [3] and [10], the technical requirements for 6G networks are stated. Table 1.1 lists a comparison of the technical requirements for 5G and 6G wireless networks [3, 9, 10].

1.1.2 Enabling Technologies

According to [4], these are the technologies that enable networks support the 5G requirements. Massive MIMO refers to the use of large-scale or multiple antenna systems at base stations to improve network capacity and increase data rate. Device-to-device (D2D) communications allow direct communication between devices. Massive machine communications aim to support the connection of large number of heterogeneous devices. Moving networks support constant communication links for moving communication devices. Ultra-dense networks enable quality communication in base-station-dense, user-dense, or interference-dense networks through proper interference and energy management. Ultra-reliable networks allow high degrees of availability. Multi-radio access technology permits devices access not only to 5G networks but also various 3G and 4G networks. Full duplex technology allows radios to use the downlink and uplink channels simultaneously which leads to increase in network capacity. mmWave communications allow the utilization of unsaturated spectrum in the GHz band which would permit larger bandwidth allocation and higher data rates. Spectrum sharing are techniques that allow access to or sharing of dedicated licensed spectrum in order to increase network capacity and reliability. Lastly, cloud technologies, such as mobile cloud computing, offer powerful computing and storage capabilities to mobile users.

The 6G enabling technologies are compiled in [3]. To increase the data rate and spectral and energy efficiency, communications above 6 GHz are being developed. While 5G focuses on mmWave communications, 6G is expected to delve beyond mmWave into the THz spectrum. 6G networks anticipate transceivers with integrated frequency bands which would allow radios to operate in microwave, mmWave, and THz bands. Large intelligent surfaces (LIS) build on massive MIMO technology by enabling communications through holographic radio frequency and holographic MIMO. Edge AI permits access to reinforcement learning algorithms to process huge amount of data and generate meaningful information. The integration of terrestrial, aerial, and satellite networks can assist in enabling ubiquitous and highly-reliable 6G networks. Energy transfer and harvesting for 6G refers to the use of backscatter as a replacement to wireless energy transfer. Finally, beyond 6G technologies such as quantum computing and communications are expected to help the implementation of 6G network security and extended coverage.

1.2 5G/6G Wireless Networks

1.2.1 Ultra-Dense Networks

Ultra-dense networks (UDNs) are networks with high cell density or number of cells per coverage area. Networks with cell density greater than 10^3 cells/km^2 are considered ultra-dense [11]. The small cells in UDN can be classified as either base stations (BSs) or access points (APs) [12]. The BSs can perform the tasks of a macrocell with lower power over a smaller coverage area. The two types of BSs in UDNs are picocells and femtocells. Picocells can be deployed outdoors and indoors by a network operator. Femtocells have lower power and smaller coverage area than a picocell, and they can be deployed indoors by users. Meanwhile, APs are installed to extend the coverage area of a macrocell. Relays and remote radio heads (RRHs) are two common ways of extending a macrocell. The difference between the two APs is that relays are installed to improve the coverage at the edge of the network, whereas RRHs are deployed to reach locations far from a BS. A typical ultra-dense network is illustrated in Fig. 1.1. The picocells and femtocells increase the capacity within a macrocell. Meanwhile, the relays and RRHs extend the macrocell coverage at the edge and at remote locations, respectively.

The main advantages of UDNs over traditional networks are high cell density, idle cell mode capabilities, advanced interference management and frequency reuse, backhaul management, and prevalent line-of-sight (LOS) transmissions [13]. High cell density enables the network to handle more active users per coverage area. As a result, the network throughput, or the average rate of data transmission per

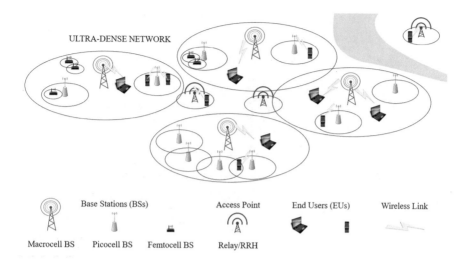

Fig. 1.1 Overview of an ultra-dense network

unit area, increases. When there are inactive cells in the network, UDNs allow these cells to operate on an idle mode. During this mode, a BS can be turned off to alleviate the inter-cell interference. Moreover, UDNs can also implement other advanced interference management and frequency reuse techniques, such as code-division multiple access (CDMA) and orthogonal frequency-division multiple access (OFDMA), to reduce the interference and increase the spectral efficiency within the network. Finally, because of the ubiquitous small cell BSs deployment in UDNs, users have a high possibility of getting a LOS transmission from the BS. Consequently, propagation models with dominant LOS component, such as Rician fading model, can be utilized to characterize and analyze UDN channels and transmissions.

Implementation of UDNs has been possible due to several enabling technologies. The authors of [13] has compiled these technologies: software-define networks (SDN), distributed antenna systems (DAS), cloud-radio access network (RAN), mmWaves networks, massive-MIMO, device-to-device (D2D) communication, multi-radio access technologies (RAT), proactive caching, and Internet-of-Things (IoT).

1.2.2 Device-to-Device Communications

Device-to-device (D2D) communication is a technology that allows nearby mobile devices to communicate directly rather than indirectly through cellular base stations or a core network [14]. D2D communication can be classified as either inband

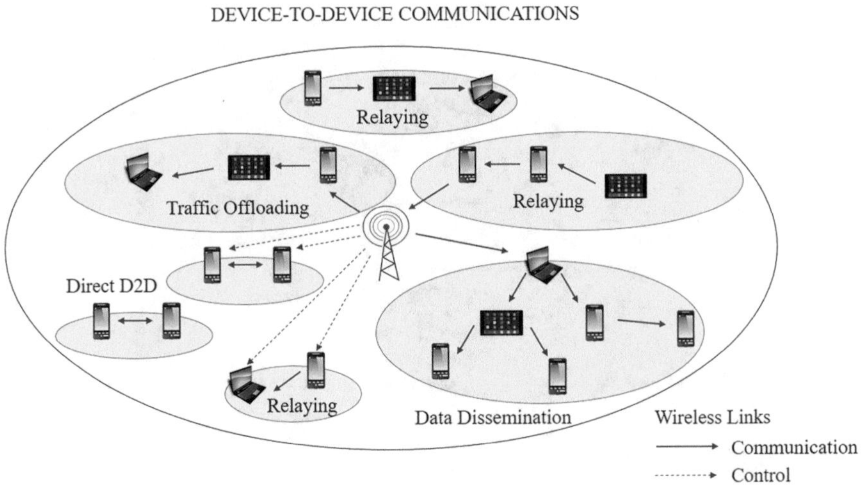

DEVICE-TO-DEVICE COMMUNICATIONS

Fig. 1.2 Several configurations of D2D communications

or outband. In inband D2D, the D2D communication is established in a licensed cellular spectrum. On the other hand, outband D2D sets up communication in an unlicensed spectrum. There are four types of D2D communications according to [15]: operator-controlled device relaying, operator-controlled direct D2D communications, device-controlled device relaying, and device-controlled direct D2D communication. The operator controls the link establishment of operator-controlled communications, whereas the source and destination devices control the link establishment of device-controlled communications. Figure 1.2 depicts various configurations of D2D communications. A direct D2D link is a direct line of communication between two D2D devices, whether its establishment is controlled by the operator or by the devices themselves. Relaying happens when a device passes data from one device to the next device. It can also be supported by the operator or by the devices themselves. Offloading and data dissemination are ways of downloading content from the network.

The advantages of D2D communications include improved spectral efficiency, increased throughput, reduced delay, improved energy efficiency, and fairness [16]. Due to the proximity of the D2D devices with each other, frequency reuse can be utilized more persistently in a coverage area, and hence, the spectral efficiency of the network is enhanced. Moreover, since nearby mobiles devices can communicate directly, higher throughput and lower delay between the devices are achievable. As a consequence, the traffic load at the backhaul network will be reduced. D2D communication also allows mobile devices to transmit at a lower power because of the proximity in distance of other D2D devices. As a result of achieving higher throughput with lower power, the energy efficiency of D2D devices is enhanced. Fairness in D2D can be achieved by obeying or satisfying the QoS requirements specified for the users.

Some of the key applications of D2D include traffic offloading, emergency services, cellular coverage extension, data dissemination, reliable health monitoring, and mobile tracking and positioning [17]. When the D2D links between mobile devices are stable, traffic offloading is a great option for mitigating the traffic congestion at cellular core networks. D2D communications can also serve as temporary networks in emergency situations where the cellular networks have limited services or are completely out-of-service. Meanwhile, D2D communications can extend the network coverage of cellular networks by improving the coverage at the edge network. For good quality and stable D2D links, data dissemination between nearby mobile devices can be achieved. In clinics and hospitals, reliable health monitoring of patients can be performed via attached D2D-enable devices from which the doctors and nurses can easily keep track of the patients' health conditions. Finally, D2D communication can be applied in mobile tracking and positioning in which deployed D2D-enabled devices can cooperate to track the position of a target or object.

1.2.3 Internet-of-Things

Internet of things (IoT) is the interconnection of large number of heterogeneous smart devices through the Internet. It allows smart devices to communicate, make decisions, and deliver services together. The IoT has six building blocks: identification, sensing, communication, computation, services, and semantics [18]. Identification refers to naming and matching services with their demand. Sensing involves IoT devices gathering data and sending them back to a data collector entity. Communication is any technology that allows the connection of heterogeneous devices to implement a number of smart applications. Computation refers to the computing ability of the IoT which is implemented through processing units and software applications. Services denote the applications offered by IoT such as smart home, intelligent transportation systems, and smart grids. Lastly, semantics depicts the ability to extract information from the data collected by the IoT devices in order to deliver the necessary services. Meanwhile, an IoT network has three main components: sensors/devices, IoT gateways, and cloud/core network [19]. The sensors/devices represent the data source or the source of measurement data in an IoT network. The IoT gateways refer the data communication network or the collectors of the measurement data from the sensors/devices for pre-processing. The cloud/core network is responsible for the data processing. The backhaul networks are responsible in forwarding the data from end devices to cloud servers. An illustration of IoT is shown in Fig. 1.3, where IoT connects heterogeneous devices belonging on each application or use-cases.

The emerging IoT applications include smart home, intelligent transportation system, smart city, industrial, and smart healthcare [20]. In order to successfully implement these applications, the key requirements include low deployment cost,

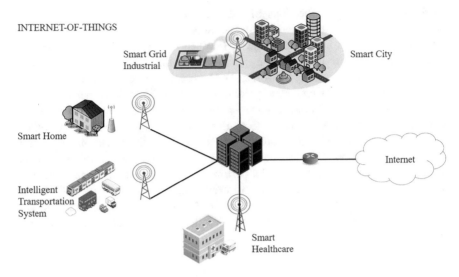

Fig. 1.3 Applications of Internet-of-Things (IoT)

long battery life, low device cost, extended coverage area, support for massive number of connected devices (scalability), and security and privacy.

1.2.4 Unmanned Aerial Vehicle Networks

Unmanned aerial vehicles (UAVs), also known as drones, are flying vehicles known for its mobility, maneuverability, and flexibility in applications. UAVs can be classified according to payload, flying mechanism, range and altitude, speed and flight time, and power supply [21]. Payload refers to the maximum weight that a UAV can carry. The equipment that the drones carry may depend on the application or mission of the UAV. For instance, base stations (BSs) or remote radio heads (RRHs) can be mounted on the UAVs that are assigned to extend the capabilities of a terrestrial communication network. UAVs can also be classified according to their flying mechanism: multi-rotor drones or rotary-wings drones can take-off (and land) vertically and hover over a location; fixed-wing drones glide through air and travel faster; and hybrid fixed/rotary wing drones are capable of operating as multi-rotor or fixed-wing drones. The functionality of UAVs are also dictated by range and altitude. The range of a UAV denotes the distance where the UAV can be controlled remotely. Typically, larger drones have longer range than smaller drones. The altitude of a UAV refers to the maximum height a UAV can travel. Low-altitude platforms (LAP) are more suitable in assisting terrestrial cellular networks, while high-altitude platforms (HAP) are currently utilized in strengthening the Internet presence around the world. The speed and flight time, or the amount of time spent on air without recharging, depend on the size of the drone. Typically, larger drones have higher speed and longer flight time than smaller drones. The power supply of UAVs depends on factors such as flight time, payload, and application. Recent power sources for UAVs are batteries, fuel, and solar energy.

UAVs have played major roles in recent advances in communications, civil, and military applications. UAVs in communication networks can be utilized either as aerial base stations to extend the coverage, increase the capacity, and improve the reliability and energy efficiency of wireless communication networks or as mobile access points to facilitate applications such as real-time video streaming and content delivery [22]. UAVs can be integrated in civil services such as search and rescue operations, remote sensing, security, and surveillance due to their ease of deployment, low maintenance cost, high-mobility, and ability to hover [23]. Furthermore, UAVs can speed-up search and rescue operations in case of natural or man-made disasters, assist in gathering and delivering data from ground sensors to base stations, and improve safety, coverage, range, robustness, and efficiency of surveillance [24].

Applications of UAVs in wireless communications have gotten tremendous attention because of the benefits and enhancements that UAVs offer to wireless networks. In [22], the authors discussed the key contributions of UAVs in wireless networks. First, UAVs are deployed as aerial base stations in 5G communications and beyond. These aerial base stations are integrated in 5G infrastructures to amplify network coverage area, improve data transmission rates, support public safety and security operations, extend terrestrial network connectivity, carry 3D MIMO and mmWave antennas to boost signal strength, assist Internet-of-Things (IoT) communications involving massive number of devices, sensors, and vehicles, and implement more efficient data caching. Second, UAVs are connected to a terrestrial cellular network as user equipment in order to facilitate applications such as package delivery and virtual reality. Third, a group of UAVs can form a flying ad-hoc network (FANET) in which the drones can communicate with each other in an ad-hoc manner and perform tasks such as relay networks and remote sensing. Other UAV applications in wireless networks include backhaul for terrestrial networks and wireless data center for smart cities. Figure 1.4 shows a general view of the role UAVs in wireless networks. The UAVs not only can act as users of terrestrial wireless networks to be able to accomplish their mission but also can act as flying BSs to extend network coverage and as FANETs to execute group missions such as data gathering, surveillance, and search and rescue operations.

1.2.5 Mobile Edge Networks

The proliferation of mobile devices has resulted in huge amount generated data and has put a lot of pressure on the wireless network infrastructures that process these data. To be able to satisfy the end users in terms of quality of experience (QoE), wireless networks must be able to deliver high quality services on time.

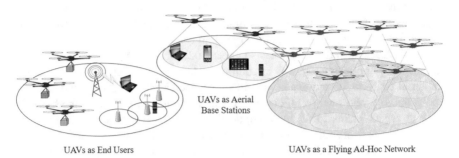

UNMANNED AERIAL VEHICLE NETWORKS

UAVs as Aerial
Base Stations

UAVs as End Users   UAVs as a Flying Ad-Hoc Network

Fig. 1.4 A general view of UAVs in wireless networks

Hence, network structures that allow high throughput and low latency services have
become a primary research topic. One of many technologies that can satisfy these
requirements is mobile edge networks (MENs). In MENs, network functions that are
usually executed at core or cloud networks are brought closer to end users through
edge networks. Therefore, MENs effectively reduce the latency due to the proximity
of the edge network to the end users while still being able to deliver high throughput
to the end users. Aside from reduced latency and high throughput, other advantages
of MENs include bandwidth reduction, high energy efficiency, and proximity and
context-aware services.

MENs have four main architectures: multi-access edge computing (MEC, and
formerly known as mobile edge computing), fog computing, cloudlet, and edge
caching [25]. MEC allows end users access to cloud computing capabilities through
MEC servers located near the end users. Fog computing is composed of fog nodes
that are highly distributed within the area of coverage, which is ideal for IoT
applications. Meanwhile, cloudlets can be deployed at WiFi access points or LTE
base stations [26]. Lastly, edge caching allows content storage at edge networks
to reduce latency associated with the request and transfer of content from the core
network. These edge network architectures are depicted in Fig. 1.5 together with a
core network.

Applications and use cases of MENs include dynamic content delivery, augmented reality, computation assistance, video streaming and analysis, IoT, connected vehicles, and wireless big data analytics [25]. Edge caching permits dynamic
content delivery, while MEC enables augmented reality or virtual reality, both
due to the decrease in latency associated with edge networks. MEC can offer
computation assistance to end users with computation-intensive tasks such as video
streaming and analysis. MEC can assist in connected vehicles road safety and
control applications. Meanwhile, the distributed architecture of fog computing
supports IoT services in healthcare, smart grid, smart homes, smart cities, and other
industries that rely on wireless sensor systems. Finally, fog computing can process
big data generated by sensors and devices scattered across a wide area.

CORE and EDGE NETWORKS

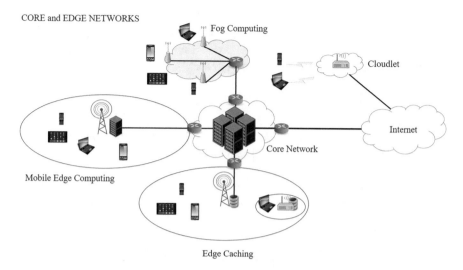

Fig. 1.5 An architecture of core and edge networks

1.3 Introduction to Mean Field Games

Mean field game (MFG) theory is introduced by Lasry and Lions in [27]. It deals with the study and analysis of differential games with infinitely many decision makers or players. It studies the solution concept of Nash equilibrium of games with very large number of indistinguishable players. The term *indistinguishable* refers to a scenario where players share common structures of the model and are allowed to have heterogeneous states [28].

Since the number of players in an MFG is large and the players are indistinguishable, the game may be taken from the point of view of a *representative* or *reference* player playing against the aggregate or collective behavior of other players in the game. Therefore, an MFG can be formulated as a couple of differential equations: a Hamilton-Jacobi-Bellman (HJB) equation characterizes the evolution of a player's optimized cost-to-go given the state distribution of all the players; and a Fokker-Planck-Kolmogorov (FPK) equation models the evolution of the state distribution of the players that are behaving optimally.

Consider a non-cooperative game with large number of indistinguishable players. Let $x = X(t)$ be the state of a player and $m = m(x, t)$ be the *mean field* or distribution of players over state x at time t. Suppose a representative player wants to minimize the cost functional J subject to the state dynamics $dX(t)$ by seeking the appropriate input or control u in the set of admissible controls $\mathcal{U}$. That is,

$$\min_{u \in \mathcal{U}} \quad J(u, m) = \mathbb{E}\left[\int_0^T r(x, u, m, t)\, dt + g(x(T), m(x, T))\right],$$

$$\text{subject to} \quad dX(t) = f(x, u, m, t)\, dt + \sigma\, dW(t),$$

$$X(0) = x_0,$$

(1.1)

where $r(x, u, m, t)$ corresponds to the running cost, $g(x(T), m(x, T))$ refers to the terminal cost or the cost at terminal time T, $f(x, u, m, t)$ denotes the average rate of change of the state, σ is the diffusion constant, and $W(t)$ is a standard Wiener process.

Define the value function or cost-to-go function as $v(x, t) = \min_{u \in \mathcal{U}} J(u, m)$. The optimal control u^* is the input or control that achieves the minimum cost $J(u^*, m)$. It satisfies the equation

$$J(u^*, m) = \mathbb{E}\left[\int_0^T r(x, u^*, m, t)\, dt + g(x(T), m(x, T))\right] = v(x, t), \qquad (1.2)$$

$\forall t \in [0, T]$. In other words, the optimal control u^* minimizes the cost functional J, and the corresponding cost is equal to the value $v(x, t)$.

It can be proved that $v(x, t)$ solves the HJB equation. Moreover, the distribution of players, whose states move according to $dX(t)$, evolves according to the FPK equation. Hence, an MFG can be expressed as a pair of HJB and FPK equations,

$$-\frac{\partial v}{\partial t}(x, t) - H(x, m, p, t) = \frac{\sigma^2}{2} \Delta_x v(x, t),$$

$$\frac{\partial m}{\partial t}(x, t) + \operatorname{div}\left(\frac{\partial H}{\partial p}(x, m, p, t) m(x, t)\right) = \frac{\sigma^2}{2} \Delta_x m(x, t),$$

(1.3)

with boundary conditions $v(x, T) = g(x)$ and $m(x, 0) = m_0(x)$. The first equation in (1.3) refers to the HJB equation that characterizes the optimized reaction of a representative player against the mean field, whereas the second equation in (1.3) denotes the FPK equation that describes the evolution of the mean field of the players that behave optimally [28].

Meanwhile, $H(x, m, p, t)$ is called the Hamiltonian, and it is defined mathematically as

$$H(x, m, p, t) = \min_{u \in \mathcal{U}} \left[r(x, u, m, t) + f(x, u, m, t) \cdot p(t)\right], \qquad (1.4)$$

where $p(t) = \nabla_x v(x, t)$.

Furthermore, the optimal control u^* is the solution to the partial differential equation

$$f(x, u^*, m, t) = \frac{\partial H}{\partial p}(x, m, p, t). \tag{1.5}$$

For a more detailed discussion, a tutorial of prerequisites and MFG is presented in Chap. 2.

1.4   Research Works on Mean Field Games in Wireless Networks

This section introduces several works on MFG applications in wireless networks considered in this book. These works include MFG applications in ultra-dense D2D networks, ultra-dense UAV networks, and user-dense MEC networks. Another research work introduced in this section is a multiple-population MFG (MPMFG) for opinion evolution in social networks. This work focuses on a recently developed MFG framework, which is an MFG with several populations and its application in analyzing the behavior of users in a multiple-population social network. The last research work introduced in this section deals with energy and time efficient computation offloading in MEC networks. The computation offloading algorithms are based on a direct method of solving an MFTG in which the MFTG solution is transformed such that solving coupled partial differential equations is avoided. For more information about other research works in the literature, a survey of applications of MFG in wireless networks is provided in Chap. 3.

1.4.1   Single-Population Mean Field Games for Ultra-Dense Networks

D2D communications provide significant performance enhancement in terms of spectrum and energy efficiency by proximity and frequency reuse. However, such performance enhancement is largely limited by mutual interference and energy availability, in particular, in ultra-dense D2D networks. In the first work featured in Chap. 4, interference dynamics and available energy of the generic device are both considered in the MFG theoretic framework with the interference mean-field approximation. In addition to the remaining energy state of the battery, the effects of the interference dynamics of the generic device to others and all others' interference dynamics introduced into the generic device are both considered for power control. Then, a novel energy and interference aware power control policy is proposed, which is a joint finite difference algorithm based on the Lax-Friedrichs scheme and the Lagrange relaxation to solve the coupled HJB and FPK equations of the corresponding MFG, respectively.

Ultra-dense UAV networks can significantly improve the system capacity and networks coverage. However, it is still a challenge to reduce interference and improve energy efficiency (EE) of UAVs. In the second work featured in Chap. 4, a downlink power control problem of maximizing the EE in an ultra-dense UAV network is investigated. The power control problem is formulated as a discrete MFG to imitate the interactions among a large number of UAVs. Then, the MFG framework is transformed into a Markov decision process (MDP) to obtain the equilibrium solution of the MFG due to the dense deployment of UAVs. Specifically, a deep reinforcement learning-based MFG (DRL-MFG) algorithm is proposed to suppress the interference and maximize the EE by using deep neural networks (DNN) to explore the optimal power strategy for UAVs.

MEC can use the distributed computing resources to serve the large numbers of mobile users in the next generation of communication systems. In this new architecture, a limited number of mobile edge servers will serve a relatively large number of mobile users. Heterogeneous servers can provide either single resource or multiple different resources to the massive number of selfish mobile users. In the third work featured in Chap. 4, two system models and problems are formulated as two non-cooperative population games to achieve high quality of service (QoS) and low latency under these two cases. Then, the proposed mean field evolutionary approach with two different strategy graphs are applied to solve the load balancing problems under those two cases.

1.4.2 Multiple-Population Mean Field Game for Social Networks

Social networks (SNs) are sets or groups of people with some patterns of contacts or interactions between them, forming meaningful social relationships [29]. These networks have evolved from simple web-based forums to ubiquitous mobile networks. As outlined in [30], the advancement of the internet has given rise to web-based social networks such as forums and chat rooms. Meanwhile, the advent of peer-to-peer networks and dedicated social websites has caused a huge number of people to use social networking through online social networks (OSNs), such as Facebook, Twitter, WeChat, and LinkedIn. OSNs allow users sharing common interests to form online connections and social communities. Moreover, the increasing capability of mobile devices has made way for the integration of social networks in mobile environments. Such networks, called mobile social networks (MSNs), allow users to access, share, and distribute data by exploiting social relations [31]. These relationships between social network users include physical contacts, financial exchanges, group participations, among others [32].

One important aspect of social networks is understanding user behavior that is essential to social network entities such as service providers and network users. In the case of OSNs, the study of user behavior is crucial to different internet entities

[33]. It helps internet service providers (ISPs) learn the evolution of traffic patterns in OSNs. Meanwhile, it provides OSN service providers knowledge about their users' behaviors toward different situations. Finally, it enhances the experience of social network users. One way of understanding the behavior of social network users is to study the evolution of belief and opinion of these users regarding a social topic or issue. Hence, many researchers have studied and developed mathematical models for belief and opinion dynamics in social networks. These models attempt to capture emerging trends caused by the interactions and decisions of the users [34].

Numerous works in the literature have focused on the application of statistical methods, optimization techniques, and game theory to belief and opinion evolution in social networks. However, as the number of decision makers grows large, the analysis of a problem becomes more complicated due to the increasing number of interactions among the decision makers. Thus, an appropriate mathematical framework is necessary to ease the analysis and the calculation of the solution to the problem.

MFG, a subclass of game theory introduced by Lasry and Lions [27], was proposed to emulate scenarios involving large number of decision makers. MFG models the interaction of a decision maker with the collective behavior of other decision makers in the game rather than with the individual behavior of each and every decision maker. Meanwhile, a new MFG framework proposed in [35] concentrates on multiple-population settings where the populations are competing and each player in a population are acting non-cooperatively. This kind of scenarios can be found in social networks. For instance, social network users can be divided according to political orientation, gender, age, and many other categories. Since users can interact with anyone belonging to any group, it would be beneficial to study how these interactions affect the belief and opinion of these users.

Inspired by this new research in MFG, Chap. 5 introduces a multiple-population MFG (MPMFG) to model the belief and opinion evolution in social networks [36]. The proposed model aims to gain information on the behavior of social network users belonging to different groups. Specifically, the proposed model can be utilized to estimate and predict how a social network group affects the belief and opinion of other groups.

1.4.3   Mean-Field-Type Game for Multi-Access Edge Computing Networks

Mobile cloud computing (MCC) offers cloud services such as computing, caching, and communications to mobile end users. These services are processed in the cloud that might be geographically located far from the end users. Consequently, MCC suffers from high latency that is not acceptable in some applications. To alleviate this problem, multi-access edge computing (MEC) has been proposed in which the cloud services are provided at the edge network located in proximity with end users. Therefore, the move of computing services from the cloud to the edge network effectively reduces latency. Aside from low latency, other benefits of MEC include proximity, high bandwidth, real-time radio network information, and location awareness [25]. MEC can be implemented through a network of computing nodes distributed over a geographic area. These computing nodes form the multi-access edge computing network (MECN) that provides computing services to end users.

Computation offloading is one of the main services provided by an MECN where an end user equipment, such as a smartphone, can offload computation-intensive tasks or portions of it to the MECN instead of performing the task locally. The decision to offload depends on factors such as latency, bandwidth, and energy consumption of the equipment. In the literature, computation offloading has been formulated as a game theoretic problem or an optimization problem with the goal of minimizing the cost incurred by a mobile device or the network subject to constraints such as computing power, latency, and bandwidth.

In Chap. 6, the idea of computation offloading is extended to offloading among computing nodes [37] and [38]. Specifically, an MECN aggregates the computation tasks from the end users and then it offloads portions of the aggregated tasks to the computing nodes. Since an MECN can be implemented through finite number of computing nodes where a computing node can have a significant effect on the utility of the network, a mean-field-type game (MFTG), a relaxed version of MFG, is utilized to analyze and model the computation offloading problem. Specifically, non-cooperative and cooperative MFTG approaches are applied to computation offloading in MECN. In these approaches, the goal of each computing node is to minimize cost by controlling its own offloading strategy subject to the state dynamics of the MECN. Each computing node does not need to know the offloading strategy of other computing nodes in order to determine its own offloading strategy. Instead, a computing node only needs to know the mean field terms that correspond to the aggregate effect of other computing nodes to the network.

1.5   Organization and Summary

This book is organized as follows. Chapter 2 provides a unified treatment of the related fields of study essential in understanding the development of the theory of mean field games (MFGs). The discussions serve as a tutorial that assists in understanding the basic concepts of optimal control theory, differential games, and MFGs. The contributions of this chapter are summarized as follows.

- Basic concepts and results from deterministic optimal control and stochastic optimal control theories are discussed, including the dynamic programming principle (DPP) and the derivation of the Hamilton-Jacobi-Bellman (HJB) equation.
- The concepts and theory from optimal control are carried over to the study of differential game, which is basically an N-player game where each player

encounters an optimal control problem. The discussion includes the mathematical framework of deterministic and stochastic differential game problems as well as a solution concept for these games called Nash equilibrium.

- The results of optimal control theory and differential games are integrated to develop and analyze the theory of MFGs. The analysis of MFGs consists of the derivation of HJB and Fokker-Planck-Kolmogorov (FPK) equations of an MFG system and an introduction of analytic and numerical methods to solve an MFG problem. Afterwards, special forms of MFG such as linear-quadratic MFG (LQMFG) and multiple-population MFG (MPMFG) are introduced.
- A relaxed version of MFG called mean-field-type game (MFTG) is presented. The discussion includes a general mathematical framework of MFTG followed by a linear-quadratic MFTG (LQMFTG) framework. Due to its special form, a system of equations to solve an LQMFTG problem is provided.

Chapter 3 covers a survey of recent research works in MFG. The survey includes MFG-based research in wireless network technologies such as ultra-dense networks (UDNs), device-to-device (D2D) communications, internet-of-things (IoT), unmanned aerial vehicles (UAVs), and mobile edge networks (MENs). The main contributions of this survey are summarized as follows.

- The discussion on each wireless technology starts with an overview of the technology which includes its advantages, uses, and applications. Afterwards, research opportunities available on each technology are listed and described. The discussion is directed at acquiring an understanding of various wireless technologies as well as the challenges that are encountered in designing and implementing these wireless networks.
- Proposed MFG-based research works in the literature that address wireless network research opportunities are introduced. The survey is aimed at providing knowledge on the current state of research on wireless technologies and imparting ideas on how MFGs are utilized to simplify and simulate wireless network environments and to upgrade the performance of wireless networks.

Chapter 4 focuses on three MFG-based research works or case studies in ultra-dense networks such as ultra-dense D2D networks, ultra-dense UAV networks, and user-dense multi-access edge computing networks (MECNs). The contributions of this chapter are summarized as follows.

- Each case study addresses a prominent research problem in UDNs and expresses how MFG is taken advantage of in each scenario. Specifically, each case study thoroughly describes the system model of the target ultra-dense network application and the research problem to be addressed. Then, the application of MFG in reformulating the problem statement and providing solution to the problem are discussed. Moreover, the advantages of the MFG-based approach are also stated.
- The case studies are intended to give ideas about the type of practical wireless engineering problems that can be stated as MFG problems and to demonstrate the advantages or effectiveness the MFG-based problem solving approach.

Chapter 5 centers on an MFG modeling of belief and opinion evolution in social networks using multiple-population MFG (MPMFG). The main contributions of this work are summarized as follows.

- The belief and opinion evolution in a social network with a large number of users is formulated as a single-population and a multiple-population MFG models. In the single-population MFG model, the users are assumed to share a common characteristic, while in the multiple-population MFG model, the social network can be divided into several populations or groups in which users in a group share a common characteristic. Also, stubborn users who are less inclined to change their initial opinions are integrated into the models.
- The classical opinion evolution models are integrated into the MFG framework. Consequently, the resulting MFG opinion models can accurately represent the behavior of rational users (i.e., the optimization of a cost function) in a social network that are affected by the opinion evolution (i.e., the state dynamic equation) of the population. In other words, the models take into account not only the opinion dynamics caused by the user's neighbors, as in the classical models, but also the behavior or preference of a user in the form of a cost function.
- The mean field distributions that correspond to the opinion evolution of the populations are calculated by extending the techniques in solving a single-population MFG to solving a multiple-population MFG. Analytical and numerical equations are provided for both single- and multiple-population MFG models. An algorithm for both MFG models is developed, implemented, and tested.
- Simulations are provided to demonstrate the practical application and purpose of the proposed MFG models. First, theoretical results from implementing the proposed models are shown. These results are important in showcasing the theoretical aspects of the MFG-based belief and opinion evolution. Then, the estimation and prediction capabilities of the proposed models are tested using a social evolution dataset [39]. Finally, performance analysis is performed in order to prove the validity and effectiveness of the proposed MFG models.

Chapter 6, the final chapter, concentrates on an MFTG approach to computation offloading in MECN. The main contributions of this work are summarized as follows.

- Computation offloading is proposed and formulated as a non-cooperative MFTG problem where each computing node minimizes its own cost function subject to the state dynamics of the network. In the non-cooperative approach, the computing nodes operate in a decentralized manner where each computing node can compute its own computation offloading strategy without full knowledge of the strategies of other computing nodes.
- A cooperative MFTG problem of computation offloading is proposed and formulated where the computing nodes jointly minimize a global cost function subject to the state dynamics of the network. In the cooperative approach, the computing nodes operate in a centralized manner where the network determines

the offloading strategy of each computing node that minimizes the cost incurred by the edge network.

- The optimal computation offloading control profile that minimizes the cost in each case is solved using a direct approach proposed in the literature. This approach does not require solving coupled partial differential equations that complicate the problem. Instead, it involves calculation of mean-field terms that represent the behavior of the entire network.
- The non-cooperative and cooperative MFTG computation offloading algorithms are designed based on the direct approach of solving MFTG computation offloading problems. The non-cooperative algorithm is implemented in a decentralized manner, while the cooperative algorithm is implemented in a centralized manner.
- Simulations are presented to demonstrate the effectiveness of the proposed MFTG-based algorithms as well as the computation offloading behavior of a computing node under varying conditions. Moreover, the proposed non-cooperative and cooperative MFTG computation offloading algorithms are compared with typical computation offloading algorithms.

References

1. Cisco, Cisco Annual Internet Report (2018–2023) White Paper, Available: https://www.cisco.com/c/en/us/solutions/collateral/executive-perspectives/annual-internet-report/white-paper-c11-741490.html, Mar 2020
2. M. Shafi, A.F. Molisch, P.J. Smith, T. Haustein, P. Zhu, P. De Silva, F. Tufvesson, A. Benjebbour, G. Wunder, 5G: a tutorial overview of standards, trials, challenges, deployment, and practice. IEEE J. Selec. Areas Commun. **35**(6), 1201–1220 (2017)
3. W. Saad, M. Bennis, M. Chen, A vision of 6G wireless systems: applications, trends, technologies, and open research problems. IEEE Netw. **34**, 134–142 (2020)
4. A. Gupta, R.K. Jha, A survey of 5G network: architecture and emerging technologies. IEEE Access **3**, 1206–1232 (2015)
5. M. Agiwal, A. Roy, N. Saxena, Next generation of 5G wireless networks: a comprehensive survey. IEEE Commun. Surv. Tutor. **18**(3), 1617–1655 (2016)
6. J.G. Andrews, S. Buzzi, W. Choi, S.V. Hanly, A. Lozano, A.C.K. Soong, J.C. Zhang, What will 5G be? IEEE J. Selec. Areas Commun. **32**(6), 1065–1082 (2014)
7. GSMA Intelligence, Understanding 5G: Perspectives on Future Technological Advancements in Mobiles, White paper (2014)
8. S. Chen, J. Zhao, The requirements, challenges, and technologies for 5G of terrestrial mobile telecommunication. IEEE Commun. Mag. **52**(5), 36–43 (2014)
9. *Minimum Requirements Related to Technical Performance for IMT-2020 Radio Interfaces(s)*, document ITU-R M.[IMT-2020.TECH PERF REQ] (2016)
10. A. Dogra, R.K. Jha, S. Jain, A survey on beyond 5G network with the advent of 6G: architecture and emerging technologies. IEEE Access **9**, 1–37 (2020)
11. M. Ding, M. López-Pérez, G. Mao, P. Wang, Z. Lin, Will the area spectral efficiency monotonically grow as small cells go dense? in *Proceedings of the IEEE Global Communications Conference (GLOBECOM)*, San Diego, CA (2015), pp. 1–7
12. D. López-Pérez, et al., Enhanced intercell interference coordination challenges in heterogeneous networks. IEEE Trans. Wireless Commun. **18**(3), 22–30 (2011)
13. M. Kamel, W. Hamouda, A. Youssef, Ultra-dense networks: a survey. IEEE Commun. Surv. Tutorials **18**(4), 2522–2545 (2016)

14. Y.-D. Lin, Y.-C. Hsu, Multihop cellular: a new architecture for wireless communications, in *Proceedings of the IEEE INFOCOM*, vol. 3 (2000), pp. 1273–1282
15. M.N. Tehrani, M. Uysal, H. Yanikomeroglu, Device-to-device communication in 5G cellular networks: challenges, solutions, and future directions. IEEE Commun. Mag. **52**, 86–92 (2014)
16. A. Asadi, Q. Wang, V. Mancuso, A survey on device-to-device communication in cellular networks. IEEE Commun. Surv. Tutor. **16**(4), 1801–1819 (2014)
17. F. Jameel, Z. Hamid, F. Jabeen, S. Zeadally, M.A. Javed, A survey of device-to-device communications: research issues and challenges. IEEE Commun. Surv. Tutor. **20**(3), 2133–2168 (2018)
18. A. Al-Fuqaha, M. Guizani, M. Mohammadi, M. Aledhari, M. Ayyash, Internet of things: a survey on enabling technologies, protocols, and applications. IEEE Commun. Surv. Tutor. **17**(4), 2347–2376 (2015)
19. W. Yu, F. Liang, X. He, W.G. Hatcher, C. Lu, J. Lin, X. Yang, A survey on the edge computing for the internet of things. IEEE Access **6**, 6900–6919 (2017)
20. G.A. Akpakwu, B.J. Silva, G.P. Hancke, A.M. Abu-Mahfouz, A survey on 5G networks for the internet of things: communication technologies and challenges. IEEE Access **6** 3619–3647 (2017)
21. A. Fotouhi, H. Qiang, M. Ding, M. Hassan, L.G. Giordano, A. Garcia-Rodriguez, J. Yuan, Survey on UAV cellular communication: practical aspects, standardization advancements, regulation, and security challenges. IEEE Commun. Surv. Tutor. **21**(4), 3417–3442 (2019)
22. M. Mozaffari, W. Saad, M. Bennis, Y.H. Nam, M. Debbah, A tutorial on UAVs for wireless networks: applications, challenges, and open problems. IEEE Commun. Surv. Tutor. **21**(3), 2334–2360 (2019)
23. S. Hayat, E. Yanmaz, R. Muzaffar, Survey on unmanned aerial vehicle networks for civil applications: a communications viewpoint. IEEE Commun. Surv. Tutor. **18**(4), 2624–2661 (2016)
24. H. Shakhatreh, A. Sawalmeh, A. Al-Fuqaha, Z. Dou, E. Almaita, I. Khalil, N.S. Othman, A. Khreishah, M. Guizani, Unmanned aerial vehicles (UAVs): a survey on civil applications and key research challenges. IEEE Access **7**, 48572–48634 (2019)
25. S. Wang, X. Zhang, Y. Zhang, L. Wang, J. Yang, W. Wang, A survey on mobile edge networks: convergence of computing, caching, and communications. IEEE Access **5**, 6757–6779 (2017)
26. Y. Gao, W. Hu, K. Ha, B. Amos, P. Pillai, M. Satyanarayanan, Are cloudlets necessary? School of Computer Science, Carnegie Mellon University, Pittsburgh, PA, USA, Technical Report CMU-CS-15–139 (2015)
27. J.M. Lasry, P.L. Lions, Mean field games. Jpn J. Math. **2**(1), 229–260 (2007)
28. D.A. Gomes, L. Nurbekyan, E.A. Pimentel, Economic models and mean-field game theory, in *30º Colóquio Brasileiro de Matemática*, Rio de Janeiro (2015)
29. M. Newman, The structure and function of complex networks. SIAM Rev. **45**(2), 167–256 (2003)
30. N. Vastardis, K. Yang, Mobile social networks: architectures, social properties, and key research challenges. IEEE Commun. Surv. Tutorials **15**(3), 1355–1370 (2013)
31. N. Kayastha, D. Niyato, P. Wang, E. Hossain, Applications, architectures, and protocol design issues for mobile social networks: a survey. Proc. IEEE **99**(12), 2130–2158 (2011)
32. S. Wasserman, F. Faust, *Social Network Analysis: Methods and Applications* (Cambridge University Press, Cambridge, 1994)
33. L. Jin, Y. Chen, T. Wang, P. Hui, A. Vasilakos, Understanding user behavior in online social networks: a survey. IEEE Commun. Mag. **51**(9), 144–150 (2013)
34. C. Chamley, A. Scaglione, L. Li, Models for the diffusion of beliefs in social networks. IEEE Signal Proce. Mag. **30**(3), 16–29 (2018)
35. A. Bensoussan, T. Huang, M. Lauriere, Mean field control and mean field game models with several populations. arXiv Optimization and Control, Available: https://arxiv.org/abs/1810.00783
36. R.A. Banez, H. Gao, L. Li, C. Yang, Z. Han, H.V. Poor, Belief and opinion evolution in social networks based on a multi-population mean field game approach, in *Proceedings of the IEEE International Conference on Communications (ICC)*, Dublin (2020), pp. 1–6

37. R.A. Banez, L. Li, C. Yang, L. Song, Z. Han, A mean-field-type game approach to computation offloading in mobile edge computing networks, in *Proceeding of the IEEE International Conference on Communications (ICC)*, Shanghai (2019), pp. 1–6
38. R.A. Banez, H. Tembine, L. Li, C. Yang, L. Song, Z. Han, H.V. Poor, Mean-field-type game-based computation offloading in multi-access edge computing networks. IEEE Trans. Wirel. Commun. **19**(12), 7825–7835 (2020)
39. A. Madan, M. Cebrian, S. Moturu, K. Farrahi, A. Pentland, Sensing the 'Health State' of a community. Pervasive Comput. **11**(4), 36–45 (2012)

Chapter 2
Introduction to Mean Field Games and Mean-Field-Type Games

The calculation of the solution or Nash equilibrium of a differential game with N players involves solving N coupled HJB equations simultaneously. It becomes more complicated as N becomes larger due to increased interactions and coupling among the players. Consequently, mean field game (MFG) has been proposed by Lasry and Lions to reformulate a differential game problem. In an MFG, the aggregate effect of all the players is considered rather than the individual effect of each player. MFGs have been applied in many applications in economics and engineering in which the number of players is large and when the players are indistinguishable yet can have heterogeneous states.

Since the number of players in an MFG is large and the players are indistinguishable, the game can be seen from the point of view of a representative or a reference player playing against the mean field or the distribution of other players. Therefore, an MFG system can be characterized by a pair of partial differential equations (PDEs): an HJB equation and a Fokker-Planck-Kolmogorov (FPK) equation. The HJB equation characterizes the evolution of the optimized objective of a player with reference to the mean field, while the FPK equation describes the evolution of the mean field of the entire population of players that behaves optimally. In contrast, a differential game is characterized by N coupled HJB equations. However, finding its solution becomes more difficult as N increases.

In this chapter, basic concepts of game theory essential in realizing the principles of MFG are established. Afterwards, optimal control theory is introduced, which is crucial in understanding the mathematics of differential games and MFG. Then, the theory of differential games is explored to further recognize the motivation for studying MFGs. Furthermore, the concepts of optimal control theory and differential games are integrated to develop and analyze the theory of MFGs. A special form of MFG, called linear-quadratic MFG (LQMFG), and an extension of MFG to multiple populations, called multiple-population MFG (MPMFG), are also introduced. In conclusion, the concept of mean-field-type games (MFTGs) is considered. An MFTG is a relaxed version of MFG in which the number of players is not necessarily

© The Editor(s) (if applicable) and The Author(s), under exclusive license to
Springer Nature Switzerland AG 2021
R. A. Banez et al., *Mean Field Game and its Applications in Wireless Networks*,
Wireless Networks, https://doi.org/10.1007/978-3-030-86905-2_2

large and the players are not necessarily indistinguishable. MFTGs have been applied in solving various engineering problems where the strong assumptions made in MFGs may not be achievable.

2.1 Introduction

To better understand the theory of mean field games (MFGs), basic concepts of game theory are introduced in this section, which include non-cooperative and cooperative games, extensive- and strategic-form games, pure and mixed strategies, and the concept of Nash equilibrium. Afterwards, the relationship of MFG with other fields of study such as optimal control theory and differential games is described.

2.1.1 Basic Concepts of Game Theory

Game theory is the study that deals with mathematical modeling and analysis of strategic interactions involving several rational decision makers or players with different goals where the decision of each player affects the outcome for all players [1]. It has been applied in modeling and analyzing real-world decision-making problems in economics, social and natural sciences, and engineering.

A game is a mathematical model of an interactive decision making situation among many players where every player strives to attain its best possible outcome while knowing that other players are striving to attain their best possible outcome as well. A game can be non-cooperative or cooperative. In a *non-cooperative game*, the players have totally or partially conflicting interests on the outcome of a decision process. Each player needs to take its decision independent of the other players given the possible choices of the other players and their effect on the player's objective or utility. Meanwhile, in a *cooperative game*, the players are allowed to form agreements among themselves that can impact the strategic choices of these players as well as their utilities [2].

A game may also be classified as either static or dynamic. In a *static game*, the players take their actions only once, independently of each other. Moreover, a static game does not have the concept of time, and the players do not have information about the decision of the other players. In a *dynamic game*, the players have information about the decision of other players and can act more than once. In addition, a dynamic game evolves or changes with time, and hence it depends on time.

2.1.1.1 Extensive-Form and Strategic-Form Games

A game may be defined in an extensive form or a strategic form. In extensive form, a game is described in terms of the rules of the game, the sequence at which the players make their moves, the information available to the players when they make their decision, the termination rules of the game, and the utility for every possible outcome of the game. An extensive-form game is illustrated in graphical form called a *game tree*. A *game tree* comprises of vertices and edges. The vertices refer to the positions in the game, while the edges represent the transition from one position to another position. The root vertex represents the starting position of the game, whereas the terminal vertices represent the terminal positions of the game. To each terminal vertex, an outcome of the game is realized when the game terminates at that terminal vertex.

Meanwhile, a game in *strategic form*, also known as *normal form*, is characterized in terms of the set of players, the set of strategies for each player, and the utilities or *payoffs* the players receive from the outcome of the game. The strategic form ignores the dynamic aspects of the game such as the order of the moves by the players, chance moves, and the informational structure of the game. A strategic-form game is defined as follows.

Definition 2.1 A game in strategic form is an ordered triple $G=(\mathcal{N}, (\mathcal{S}_i)_{i\in\mathcal{N}}, (r_i)_{i\in\mathcal{N}})$ with the following components:

i. $\mathcal{N} = \{1, 2, \ldots, N\}$ is a finite set of players;
ii. $\mathcal{S}_i$ is the set of strategies of player i, for every player $i \in \mathcal{N}$. The set of all vectors of strategies is denoted as $\mathcal{S} = \mathcal{S}_1 \times \cdots \times \mathcal{S}_N$; and
iii. $r_i : \mathcal{S} \to \mathbb{R}$ is a utility function associating each vector of strategies $s = (s_i)_{i\in\mathcal{N}} \in \mathcal{S}$ with the utility $r_i(s)$ to player i, for every player $i \in \mathcal{N}$.

The utilities of the players are quantitatively represented by a utility function. A utility function, also known as payoff function, quantifies a players' preferences of possible outcomes of the game. It provides the outcomes of a game with corresponding payoffs or costs. For instance, the most preferred outcomes equate to the greatest payoffs or least costs, while the least preferred outcomes equate to the least payoff or greatest cost.

2.1.1.2 Pure Strategies and Mixed Strategies

A strategy of a player is a function that determines the action that the player will take at any stage of the game considering the information available to the player. A strategy may be pure strategy or mixed strategy. A pure strategy refers to the strategy that is chosen with a probability of 1 considering the information available to the player, while a mixed strategy is a probability distribution over the set of pure strategies. That is, given a set of information available, a player with mixed strategies chooses its strategy based on the probability distribution over its pure strategies. A mixed strategy can be denoted as the strategy profile $\sigma_i = (\sigma_i(s_i))_{s_i \in \mathscr{S}_i}$, where $\sigma_i(s_i)$ refers to the probability that player i plays the pure strategy s_i. The following definition provides a mathematical description of a mixed strategy and the set of mixed strategies of player i.

Definition 2.2 Let $G = (\mathscr{N}, (\mathscr{S}_i)_{i \in \mathscr{N}}, (r_i)_{i \in \mathscr{N}})$ be a strategic-form game with finite set of strategies for each player i. A mixed strategy of player i is the probability distribution over its set of strategies $\mathscr{S}_i$. Moreover, the set of mixed strategies of player i is denoted by

$$\Sigma_i = \left\{ \sigma_i : \mathscr{S}_i \to [0, 1] : \sum_{s_i \in \mathscr{S}_i} \sigma_i(s_i) = 1 \right\},$$

for every player $i \in \mathscr{N}$.

In words, the set Σ_i contains all the possible probability distributions assigned by player i over its strategy set $\mathscr{S}_i$. The extension of a strategic-form game in mixed strategies is defined as follows.

Definition 2.3 Let $G = (\mathscr{N}, (\mathscr{S}_i)_{i \in \mathscr{N}}, (r_i)_{i \in \mathscr{N}})$ be a strategic-form game for every player $i \in \mathscr{N}$. Also, assume that the set of pure strategies $\mathscr{S}_i$ is nonempty and finite. Let $\mathscr{S} = \mathscr{S}_1 \times \cdots \times \mathscr{S}_N$ be the set of pure strategy profile. The mixed-strategy extension of G is the game

$$\Gamma = (\mathscr{N}, (\Sigma_i)_{i \in \mathscr{N}}, (R_i)_{i \in \mathscr{N}}),$$

where, for each player $i \in \mathscr{N}$, Σ_i is the set of strategies and $R_i : \Sigma \to \mathbb{R}$ is the utility or payoff function that associates each strategy profile $\sigma = (\sigma_1, \ldots, \sigma_N) \in \Sigma = \Sigma_1 \times \cdots \times \Sigma_N$ with the payoff

$$R_i(\sigma) = \mathbb{E}_\sigma[r_i(\sigma)] = \sum_{(s_1, \ldots, s_N) \in \mathscr{S}} r_i(s_1, \ldots, s_n) \sigma_1(s_1) \sigma_2(s_2) \cdots \sigma_N(s_N).$$

This extension of strategic form game to mixed strategies covers the set of finite players $\mathscr{N}$, and for each player $i \in \mathscr{N}$, the set of mixed strategies Σ_i, and the overall utility or payoff $R_i(\sigma)$ taken as the mathematical expectation $\mathbb{E}[\cdot]$ of the utility r_i over the mixed strategy profile $\sigma \in \Sigma$.

2.1.1.3 Nash Equilibrium

The outcome or set of outcomes that will result given the behavior of the players refers to a solution of the game. A stable solution of the game is also called an equilibrium point. The Nash equilibrium, defined by John Nash, is an equilibrium point of the game referring to a strategy vector or profile $s^* = (s_1^*, \ldots, s_n^*)$ at which no player $i \in \mathscr{N}$ has a profitable deviation from s_i^*.

Definition 2.4 A strategy profile $s^* = (s_1^*, \ldots, s_N^*)$ is a Nash equilibrium if, for each player $i \in \mathcal{N}$ and each strategy $s_i \in \mathcal{S}_i$, the inequality,

$$r_i(s^*) \geq r_i(s_i, s_{-i}^*),$$

is satisfied, where s_{-i}^* is the strategy profile containing the strategies of all players except player i. The utility profile $r(s^*) = (r_i(s^*))_{i \in \mathcal{N}}$ is the equilibrium payoff corresponding to the Nash equilibrium s^*.

Another way of defining a Nash equilibrium is through the concept of best response strategy.

Definition 2.5 The best response strategy of player i to strategy profile s_{-i} satisfies

$$r_i(s_i, s_{-i}) = \max_{t_i \in \mathcal{S}_i} r_i(t_i, s_{-i}).$$

That is, the best response of a player i implies that, if each of the other players adheres to s_{-i}, then player i cannot do better than playing its best response strategy. Consequently, a Nash equilibrium in terms of the best response strategy is given as follows.

Definition 2.6 The strategy profile $s^* = (s_1^*, \ldots, s_N^*)$ is a Nash equilibrium if s_i^* is a best response strategy to s_{-i}^* for every player $i \in \mathcal{N}$. That is,

$$r_i(s_i^*, s_{-i}^*) \geq r_i(s_i, s_{-i}^*),$$

for all $s_i \in \mathcal{S}_i$.

In other words, in a Nash equilibrium, no player would have an incentive to deviate from s_i^* given the best response strategies s_{-i}^* of other players. The definition of Nash equilibrium can be extended to games involving mixed strategies.

Definition 2.7 A mixed strategy profile $\sigma^* = (\sigma_i^*, \sigma_{-i}^*) \in \Sigma$ is a mixed strategy Nash equilibrium if, for each player $i \in \mathcal{N}$ and each mixed strategy $\sigma_i \in \Sigma_i$, the inequality,

$$r_i(\sigma_i^*, \sigma_{-i}^*) \geq r_i(\sigma_i, \sigma_{-i}^*),$$

is satisfied, where σ_{-i}^* is the mixed strategy profile containing the mixed strategies of all players except player i. The utility profile $r(\sigma^*) = (r_i(\sigma^*))_{i \in \mathcal{N}}$ is the equilibrium payoff corresponding to the Nash equilibrium σ^*.

Moreover, since $r_i(\sigma_i^*, \sigma_{-i}^*) = \sum_{s_i \in \mathcal{S}_i} \sigma_i(s_i) r_i(s_i, \sigma_{-i}^*)$, it is sufficient to check only the deviations in pure strategies when determining whether a mixed strategy profile is a Nash equilibrium or not. Hence, a mixed strategy profile is a mixed strategy Nash equilibrium if, for each player $i \in \mathcal{N}$ and for each mixed strategy

$\sigma_i \in \Sigma_i$, the inequality

$$r_i(\sigma_i^*, \sigma_{-i}^*) \geq r_i(s_i, \sigma_{-i}^*),$$

is satisfied.

2.1.2   Mean Field Games and Related Fields of Study

To understand the theory of MFGs, knowledge of related fields of study such as optimization theory, game theory, optimal control theory, and differential games are necessary.

Optimization theory studies techniques in solving optimization problems. In an optimization problem, the goal of the decision maker is to find the optimal value of an optimization variable that optimizes an objective function (i.e., maximizes a payoff function or minimizes a cost function) subject to constraints on the values of the optimization variables.

Let u be an optimization variable with values contained in a set $\mathcal{U}$. Suppose J is the objective or cost function used to evaluate the performance of u. An optimization problem in which the decision maker seeks the value of u that minimizes the cost function J may be written as

$$J(u^*) = \min_{u \in \mathcal{U}} J(u),$$

where u^* refers to the optimal value of u and $J(u^*)$ denotes the minimum cost based on u^*.

Optimal control theory is the study of optimal control problems or continuous-time dynamic optimization problems with differential equation constraints. In an optimal control problem, the decision maker seeks the value of $u \in \mathcal{U}$ that optimizes the objective function J subject to the differential equation constraint $dx(t)$. Mathematically, an optimal control problem can be written as

$$J(u^*) = \min_{u \in \mathcal{U}} J(u),$$

$$\text{subject to } dx(t) = f(x, u, t)\, dt,$$

where u^* denotes the optimal value of u and $J(u^*)$ denotes the minimum cost based on u^*.

Meanwhile, game theory deals with strategic interactions involving several rational decision makers. Let $\mathcal{N}$ be the set of N decision makers. Each decision maker $i \in \mathcal{N}$ wants to find the optimal value of $u_i \in \mathcal{U}_i$ that optimizes its objective function J_i based on the action of other decision makers, $u_{-i} = (u_1, \ldots, u_{i-1}, u_{i+1}, \ldots, u_N)$. A game theoretic problem may be written as

$$J_i(u_i^*, u_{-i}) = \min_{u_i \in \mathcal{U}_i} J_i(u_i, u_{-i})$$

for all $i \in \mathcal{N}$, where u_i^* denotes the optimal value of u_i and $J_i(u_i^*, u_{-i})$ refers to the minimum cost based on u_i^* and u_{-i}.

Differential games extend optimal control theory to several decision makers. Let $\mathcal{N}$ be the set of N decision makers. Each decision maker $i \in \mathcal{N}$ wants to find the optimal value of $u_i \in \mathcal{U}_i$ that optimizes its objective function J_i that takes into account the action of other decision makers, $u_{-i} = (u_1, \ldots, u_{i-1}, u_{i+1}, \ldots, u_N)$, subject to the differential equation constraint $dx(t)$. A differential game problem may be written as

$$J_i(u_i^*, u_{-i}) = \min_{u_i \in \mathcal{U}_i} J_i(u_i, u_{-i}),$$

$$\text{subject to } dx(t) = f_i(x, u_i, u_{-i}, t) \, dt,$$

for all $i \in \mathcal{N}$, where u_i^* denotes the optimal value of u_i and $J_i(u_i^*, u_{-i})$ denotes the minimum cost based on u_i^* and u_{-i}.

An MFG is an integration of all these concepts. Let $\mathcal{N}$ be the set of N decision makers. Assume that N is large, and the decision makers are symmetric and indistinguishable. Any decision maker wants to find the optimal value of $u \in \mathcal{U}$ that optimizes its objective function J that takes into account the mean field m that captures the collective behavior of other decision makers subject to the differential equation constraint $dx(t)$. An MFG problem may be written as

$$J(u^*, m) = \min_{u \in \mathcal{U}} J(u, m),$$

$$\text{subject to } dx(t) = f(x, u, m, t) \, dt,$$

where u^* denotes the optimal value of u and $J(u^*, m)$ denotes the minimum cost based on u^* and m. Note that in an MFG, the decision of other decision makers u_{-i} has been replaced by the mean field m. Since the decision makers are symmetric and indistinguishable, an MFG problem can be taken from the point of view of any representative decision maker. Therefore, an MFG problem can be regarded as an optimal control problem in which given the mean field m, a representative decision maker aims to optimize its objective function J subject to the state dynamic equation $dx(t)$ by finding the optimal value of u. Figure 2.1 illustrates the relation of MFG with other related fields of study.

In an MFG, the number of players is large, and each player aims to optimize its own utility according to the states of the other players in the group. In a dynamic system where the state or behavior of each player changes with time, the distribution of the states, or the mean field, evolves as well. Meanwhile, when the players react to the mean field, the optimized utility of a player, or the value function, varies accordingly. If each player is behaving optimally given the collective behavior of the other players, the dynamics, or the evolution with respect to time, of the value function and the mean field can be approximated by the mean field game system.

The mean field game system consists of a Hamilton-Jacobi-Bellman (HJB) equation for the value function $v(x, t)$ and a Fokker-Planck-Kolmogorov (FPK)

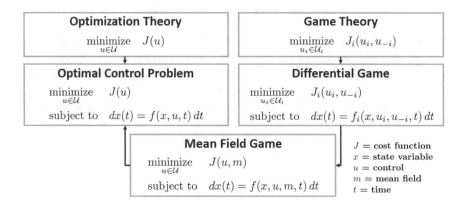

Fig. 2.1 Mean field game and related fields of study

equation for the mean field density $m(x, t)$,

$$-\frac{\partial v}{\partial t}(x, t) - H(x, m, \nabla_x v, t) = \nu \Delta_x v(x, t),$$

$$\frac{\partial m}{\partial t}(x, t) + \text{div}\big[f(x, u, m, t)m(x, t) \big] = \nu \Delta_x m(x, t),$$

(2.1)

with boundary conditions $m(x, 0) = m_0(x)$ and $v(x, T) = g(x)$ and a non-negative parameter ν. The function H is called the Hamiltonian, which is optimized by varying the control variable u. Meanwhile, div refers to the divergence operation, ∇_x denotes the gradient operation, and Δ_x refers to the Laplace operation.

The first equation in (2.1) refers to the HJB equation that corresponds to the evolution of the solution to the optimal control problem that each player is solving. The second equation in (2.1) refers to the FPK equation that characterizes the evolution of the distribution of the players given that each player is behaving optimally.

In the following sections, mathematical fields of study that help in developing and understanding MFGs are discussed. These fields of study are optimal control theory and differential games.

2.2 Optimal Control Theory

Optimal control theory is the study of optimal control problems or continuous-time dynamic optimization problems with differential equation constraints. In an optimal control problem, a single decision maker (e.g., an individual player or the system itself) decides which input or control function optimizes the objective function subject to the state dynamics. Optimal control theory provides mathematical techniques to find the control function of a dynamical system that leads to an optimized objective function.

In this section, two subclasses of optimal control theory are discussed: deterministic optimal control and stochastic optimal control. The type of optimal control problem depends on the characteristic of the state of a dynamical system. A state refers to a system variable that represents a measurable quantity that evolves over time. In deterministic optimal control, the state of the system is a deterministic process in which there is no randomness involved in determining the state at any given time. In stochastic optimal control, the state of the system is a stochastic process where the state at any given time has inherent random characteristic.

The discussion of optimal control theory in this section focuses on the main concepts and results that are important in understanding MFGs. For more rigorous treatment of optimal control theory, the reader may to refer to [3, 4].

2.2.1 Deterministic Optimal Control

Let $f(x, u, \tau) : \mathbb{R}^n \times \mathbb{R}^m \times [t, T] \to \mathbb{R}^n$ be a function referring to the instantaneous rate of change of x with respect to τ. The state x of a system at time $\tau \in [t, T]$ is denoted by the mapping $x(\tau) : [t, T] \to \mathbb{R}^n$. Assume the state evolves according to a differential equation

$$dx(\tau) = f(x(\tau), u(\tau), \tau) \, d\tau,$$
$$x(t) = x,$$

(2.2)

where $x \in \mathbb{R}^n$ is the fixed initial state, $t \geq 0$ is the initial time, and T is the final or terminal time. The control $u(\tau) : [t, T] \to \mathbb{R}^m$ is a measurable function with values from a compact subset U of $\mathbb{R}^m$. It belongs to the set of all admissible controls $\mathcal{U} = \{u(\tau) : [t, T] \to U\}$.

Let $r(x, u, \tau) : \mathbb{R}^n \times \mathbb{R}^m \times [t, T] \to \mathbb{R}$ be the running cost function and $g(x) : \mathbb{R}^n \to \mathbb{R}$ be the terminal cost function or the cost that depends on the state at terminal time. The control is evaluated based on the performance metric given by a functional of the form

$$J(u) = \left[\int_t^T r(x(\tau), u(\tau), \tau) \, d\tau + g(x(T)) \right].$$

(2.3)

Given the initial state $x(t) = x$, a deterministic optimal control problem is finding an optimal control $u^*(\cdot)$ that minimizes the cost functional J,

$$J(u^*) = \min_{u(\cdot) \in \mathcal{U}} J(u) = \min_{u(\cdot) \in \mathcal{U}} \left[\int_t^T r(x(\tau), u(\tau), \tau) \, d\tau + g(x(T)) \right],$$

(2.4)

subject to (2.2). In other words, given the initial state of the system, the goal is to control the system such that the total cost is minimized from time $\tau = t$ to $\tau = T$.

The cost of implementing an optimal control is called the value function $v(x, t)$ or the cost-to-go function,

$$v(x, t) = \min_{u(\cdot) \in \mathscr{U}} J(u) = \min_{u(\cdot) \in \mathscr{U}} \left[\int_t^T r(x(\tau), u(\tau), \tau) \, d\tau + g(x(T)) \right]. \qquad (2.5)$$

Note that the value function at terminal time T is $v(x, T) = g(x)$. Furthermore, the optimal control $u^*(\cdot)$ satisfies

$$J(u^*) = \int_t^T r(x(\tau), u^*(\tau), \tau) \, d\tau + g(x(T)) = v(x, t). \qquad (2.6)$$

According to the Bellman principle of optimality, "[an] optimal policy has the property that whatever the initial state and initial decision are, the remaining decisions must constitute an optimal policy with regard to the state resulting from the first decision" [5]. Therefore, the value function $v(x, t)$, and hence the optimization problem $\min_{u(\cdot) \in \mathscr{U}} J(u)$, can be solved by realizing the Bellman principle of optimality. The dynamic programming principle (DPP) is based on this optimality principle, and it can be utilized to rewrite an optimization problem into more manageable subproblems.

2.2.1.1 Dynamic Programming Principle

The DPP takes advantage of the Bellman principle of optimality by providing a method of solving complicated optimization problems by breaking them down into simpler subproblems.

DPP is a mathematical tool used in determining the value function $v(x, t)$ at initial time t. It states that given a value function $v(x_s, s)$ where $x_s = x(s)$, the value function $v(x, t)$ at time $t < s$ can be determined by optimizing the total cost functional from time t to time s. Thus, (2.5) can be written as

$$v(x, t) = \min_{u(\cdot) \in \mathscr{U}} \left[\int_t^s r(x(\tau), u(\tau), \tau) \, d\tau + v(x_s, s) \right]. \qquad (2.7)$$

That is, if a value function $v(x_s, s)$ is known, where $x_s = x(s)$, the value function $v(x, t)$ in state x at time $t < s$ can be determined backwards using the DPP.

Theorem 2.1 *Let $v(x, t)$ be the value function at initial time t,*

$$v(x, t) = \min_{u(\cdot) \in \mathscr{U}} \left[\int_t^T r(x(\tau), u(\tau), \tau) \, d\tau + g(x(T)) \right], \qquad (2.8)$$

and $v(x, T) = g(x)$ be the value function at terminal time T.

The dynamic programming principles states that, for any time s that satisfies $t < s \leq T$, *the value function in state x at initial time t can be calculated backwards from* $v(x_s, s)$,

$$v(x, t) = \min_{u(\cdot) \in \mathcal{U}} \left[\int_t^s r(x(\tau), u(\tau), \tau) \, d\tau + v(x_s, s) \right], \tag{2.9}$$

where $x_s = x(s)$.

In other words, DPP can find the value function $v(x, t)$ for $t < s$ given the value function $v(x_s, s)$ at time s, without any information beyond time s (i.e., on time interval $[s, T]$).

2.2.1.2 Hamilton-Jacobi-Bellman Equation

An approach to solve the value function $v(x, t)$ in (2.8) is to find a partial differential equation (PDE) whose solution is the value function itself. A complication arises when $v(x, t)$ may not be differentiable, and the PDE that can be solved by $v(x, t)$ does not exist. However, dividing the time horizon $[t, T]$ into smaller subintervals of length δ and applying the DPP in (2.9) on each subinterval leads to a PDE that is satisfied by $v(x, t)$.

The Hamilton-Jacobi-Bellman (HJB) equation is a PDE whose viscosity (i.e., weaker) solution is the value function $v(x, t)$. The derivation of the HJB equation starts with rewriting (2.9) by setting $s = t + \delta$,

$$v(x, t) = \min_{u(\cdot) \in \mathcal{U}} \left[\int_t^{t+\delta} r(x(\tau), u(\tau), \tau) \, d\tau + v(x(t + \delta), t + \delta) \right], \tag{2.10}$$

where $0 < \delta < T - t$.

Assume v is continuously differentiable with respect to its arguments x and t. Expanding $v(x(t + \delta), t + \delta)$ using the Taylor series about the point $(x(t), t)$ results in

$$v(x(t + \delta), t + \delta) = v(x(t), t) + \delta \frac{\partial v}{\partial t}(x(t), t) + [x(t + \delta) - x(t)] \cdot \nabla_x v(x(t), t) + o(\delta),$$

where $\nabla_x v$ is the gradient of v with respect to x, which is an n-element column vector that contains $\partial v / \partial x_i, i = 1, 2, \ldots, n$, and $o(\delta)$ denotes the higher-order terms of the expansion. From (2.2), $x(t + \delta) - x(t)$ is equivalent to

$$x(t + \delta) - x(t) = \int_t^{t+\delta} f(x(\tau), u(\tau), \tau) \, d\tau.$$

Hence, $v(x(t + \delta), t + \delta)$ can be rewritten as

$$v(x(t + \delta), t + \delta) = v(x(t), t) + \delta \frac{\partial v}{\partial t}(x(t), t)$$

$$+ \left(\int_t^{t+\delta} f(x(\tau), u(\tau), \tau) \, d\tau \right) \cdot \nabla_x v(x(t), t) + o(\delta).$$

Substituting this equation into (2.10) yields

$$v(x, t) = \min_{u(\cdot) \in \mathcal{U}} \left[\int_t^{t+\delta} r(x(\tau), u(\tau), \tau) \, d\tau + v(x(t), t) + \delta \frac{\partial v}{\partial t}(x(t), t) \right.$$

$$\left. + \left(\int_t^{t+\delta} f(x(\tau), u(\tau), \tau) \, d\tau \right) \cdot \nabla_x v(x(t), t) + o(\delta) \right].$$

The terms $v(x, t)$, $\delta \frac{\partial v}{\partial t}(x(t), t)$, and $o(\delta)$ can be taken outside of the min because they do not depend on the control $u(t)$. Cancelling $v(x, t)$ on both sides and dividing the result by δ yield

$$\frac{\partial v}{\partial t}(x(t), t) + \min_{u(\cdot) \in \mathcal{U}} \left[\frac{\int_t^{t+\delta} r(x(\tau), u(\tau), \tau) \, d\tau}{\delta} \right.$$

$$\left. + \frac{\int_t^{t+\delta} f(x(\tau), u(\tau), \tau) \cdot \nabla_x v(x(t), t) \, d\tau}{\delta} \right] + \frac{o(\delta)}{\delta} = 0.$$

Getting the limit as $\delta \to 0$, taking into account that

$$\lim_{\delta \to 0} \frac{\int_t^{t+\delta} \phi(x(\tau), u(\tau), \tau) \, d\tau}{\delta} = \phi(x(t), u(t), t),$$

for a continuous function ϕ,

$$\lim_{\delta \to 0} \frac{o(\delta)}{\delta} = 0,$$

and consequently, acknowledging that $x(t) = x$ and $u(t) = u$ lead to the HJB equation,

$$\frac{\partial v}{\partial t}(x, t) + \min_{u \in U}[r(x, u, t) + f(x, u, t) \cdot \nabla_x v(x, t)] = 0, \qquad (2.11)$$

with terminal condition $v(x, T) = g(x)$.

The second term in (2.11) is known as the Hamiltonian H,

$$H(x, p, t) = \min_{u \in U}[r(x, u, t) + f(x, u, t) \cdot p(t)], \qquad (2.12)$$

where $p(t) = \nabla_x v(x, t)$. When the optimal control u^* is realized, the Hamiltonian can be expressed as

$$\Phi(x, p, t) = r(x, u^*, t) + f(x, u^*, t) \cdot p(t). \tag{2.13}$$

Hence, the HJB equation (2.11) becomes

$$\frac{\partial v}{\partial t}(x, t) + \Phi(x, p, t) = 0. \tag{2.14}$$

Suppose the value function $v(x, t)$ is differentiable, and it solves (2.11) and (2.12). For any time $\tau > t$, the optimal control $u^*(\tau)$ is computed using the equation

$$u^*(\tau) = \arg\min_{u \in U}[r(x^*(\tau), u, \tau) + f(x^*(\tau), u, \tau) \cdot \nabla v(x^*(\tau), \tau)], \tag{2.15}$$

and the corresponding optimal trajectory $x^*(\tau)$ solves the differential equation given by

$$dx^*(\tau) = f(x^*(\tau), u^*(\tau), \tau) d\tau, \tag{2.16}$$

with initial condition $x^*(t) = x$.

2.2.2   Stochastic Optimal Control

In a stochastic optimal control problem, the state of the system is a stochastic process. Consequently, the control will also be stochastic since the system is designed to control its stochastic state.

In the following subsections, basic concepts of stochastic processes and stochastic differential equations are reviewed. Then, Ito stochastic differentiation rule, a differentiation rule involving stochastic processes, is introduced. Moreover, the DPP for transforming a stochastic optimal control problem into a dynamic programming problem is discussed. Finally, the HJB equation for stochastic optimal control is derived.

2.2.2.1   Stochastic Process and Stochastic Differential Equations

A probability space is a mathematical space in which random variables and stochastic processes are defined. It consists of a set of all possible outcomes, a set of events where an event is a subset of outcomes, and a probability function that assigns a probability to an event.

Definition 2.8 A probability space is a triple $(\Omega, \mathscr{F}, P)$ with the following components:

i. Ω is a sample space which is a set of all possible outcomes;
ii. $\mathscr{F}$ is an event space which is a σ-algebra of subsets of Ω; and
iii. P is a probability function on the set Ω such that $P : \mathscr{F} \to [0, 1]$.

A real-valued random variable X associated with a probability space $(\Omega, \mathscr{F}, P)$ is a function that assigns a real number $X(\omega) \in \mathbb{R}^n$ to an outcome $\omega \in \Omega$. It is represented by the mapping $X(t) : \Omega \to \mathbb{R}^n$ such that for all $\alpha \in \mathbb{R}^n$, $\{\omega | X(\omega) \le \alpha\} \in \mathscr{F}$.

A stochastic process $X(t)$ is a collection of random variables $X(t), 0 \le t < \infty$, each defined on the same probability space $\Omega, \mathscr{F}, P$.

A real-valued stochastic process $W(t), t \ge 0$ is called a Wiener process, or Brownian motion, if it has the following characteristics:

i. $W(0) = 0$;
ii. $W(t)$ is continuous;
iii. $W(t)$ is Gaussian with mean $\mu = 0$ and variance $\sigma^2 = t$ (i.e., $W(t) \sim \mathscr{N}(0, t)$); and
iv. $W(t)$ has independent increments which means $W(t_1), W(t_2) - W(t_1), \ldots,$ $W(t_k) - W(t_{k-1})$ are independent random variables, where $0 < t_1 < t_2 < \cdots < t_k$.

A stochastic differential equation (SDE) is a differential equation involving the derivative of a stochastic process. It is generally defined as

$$dX(t) = f(X(t), t)\, dt + \sigma(X(t), t)\, dW(t),$$
$$X(0) = x_0, \tag{2.17}$$

for $t > 0$, where the x_0 is an initial condition and $W(t) \in \mathbb{R}^l$ is a standard Wiener process. The function $f(x, t) : \mathbb{R}^n \times [0, T] \to \mathbb{R}^n$ is called the drift, while the function $\sigma(x, t) : \mathbb{R}^n \times [0, T] \to \mathbb{R}^{n \times l}$ is called the diffusion, where $x = X(t)$.

The solution of an SDE is a stochastic process. A stochastic process $X(\cdot)$ is a solution of the SDE in (2.17) if, for all $t \ge 0$,

$$X(t) = x_0 + \int_0^t f(X(\tau), \tau)\, d\tau + \int_0^t \sigma(X(\tau), \tau)\, dW(\tau). \tag{2.18}$$

2.2.2.2 Ito Stochastic Differentiation Rule

An important tool in deriving the HJB equation for a stochastic optimal control problem is the Ito stochastic differentiation rule. It states that for an Ito drift-diffusion process,

$$dX(t) = f(t) \, dt + \sigma(t) \, dW(t),$$

$$X(0) = x_0,$$

where $X(t) \in \mathbb{R}^{n \times 1}$, $f(t) \in \mathbb{R}^{n \times 1}$, $\sigma(t) \in \mathbb{R}^{n \times l}$ and $W(t) \in \mathbb{R}^{l \times 1}$, the Ito stochastic differentiation rule for a twice-differentiable scalar function $\phi(X(t), t)$ of two real variables $X(t)$ and t is stated as

$$d\phi(X(t), t) = \left(\frac{\partial \phi}{\partial t} + f \cdot \nabla_x \phi + \frac{1}{2} \mathrm{Tr}[\sigma^\top \mathbb{H}_x(\phi) \sigma] \right) dt + (\nabla_x \phi)^\top \sigma \, dW(t),$$

(2.19)

where $\nabla_x \phi \in \mathbb{R}^{n \times 1}$ is the gradient of ϕ with respect to x, $\mathbb{H}_x(\phi) \in \mathbb{R}^{n \times n}$ is the Hessian matrix of ϕ with respect to x, $\cdot$ denotes dot product operation, Tr is the matrix trace operator, and $^\top$ is the matrix transpose operator. The stochastic process $\phi(X(t), t)$ is also an Ito drift-diffusion process.

2.2.2.3 Stochastic Optimal Control Problem

Let the state of the system be a stochastic process $X(\tau) : [t, T] \times \Omega \to \mathbb{R}^n$ which satisfies the stochastic differential equation

$$dX(\tau) = f(X(\tau), U(\tau), \tau) \, d\tau + \sigma(X(\tau), U(\tau), \tau) \, dW(\tau),$$

$$X(t) = x,$$

(2.20)

where $x \in \mathbb{R}^n$ is the initial state, $t > 0$ is the initial time, and T is the final or terminal time. The term $f(x, u, \tau) : \mathbb{R}^n \times \mathbb{R}^m \times [t, T] \to \mathbb{R}^n$ is the drift function, $\sigma(x, u, \tau) : \mathbb{R}^n \times \mathbb{R}^m \times [t, T] \to \mathbb{R}^{n \times l}$ is the diffusion function, and $W(\tau)$ is an l-dimensional Wiener process defined a probability space $(\Omega, \mathscr{F}, \{\mathscr{F}_\tau\}_{\tau \geq t}, P)$ and σ is an $n \times l$ matrix.

Meanwhile, the control process $U(\tau) : [t, T] \times \Omega \to \mathbb{R}^m$ is a measurable function with values from a compact subset $\mathbb{U}$ of $\mathbb{R}^m$. It is adapted to the filtration $\{\mathscr{F}_\tau\}_{\tau \geq t}$ which means the control is based on information known up to the present time. It belongs to the set of admissible controls $\mathscr{U} = \{U(\tau) : [t, T] \times \Omega \to \mathbb{U}\}$.

Let $r(x, u, \tau) : \mathbb{R}^n \times \mathbb{R}^m \times [t, T] \to \mathbb{R}$ be the running cost function and $g(x) : \mathbb{R}^n \to \mathbb{R}$ be the terminal cost function. The stochastic optimal control problem is finding the optimal control U^* that solves

$$J(U^*) = \min_{U(\cdot) \in \mathscr{U}} J(U) = \min_{U(\cdot) \in \mathscr{U}} \mathbb{E} \left[\int_t^T r(X(\tau), U(\tau), \tau) \, d\tau + g(X(T)) \right],$$

(2.21)

subject to (2.20). Because the state is a stochastic process, the cost becomes a random variable. Consequently, the criterion in computing the control requires the mathematical expectation of the cost functional J.

In terms of the value function $v(x, t)$, or the cost-to-go function, the optimal problem can be written as

$$v(x, t) = \min_{U(\cdot) \in \mathscr{U}} J(U) = \min_{U(\cdot) \in \mathscr{U}} \mathbb{E}\left[\int_t^T r(X(\tau), U(\tau), \tau) \, d\tau + g(X(T))\right],$$

(2.22)

where $X(t) = x$. The optimal control U^* satisfies

$$J(U^*) = \mathbb{E}\left[\int_t^T r(X(\tau), U^*(\tau), \tau) \, d\tau + g(X(T))\right],$$

(2.23)

$$= v(x, t).$$

2.2.2.4 Dynamic Programming Principle

In order to derive the HJB equation that is satisfied by $v(x, t)$ in a stochastic optimal control problem, the DPP for stochastic control is utilized. The DPP takes advantage of the Bellman principle of optimality [5], similar to the method used in the previous section for deterministic optimal control. The DPP for stochastic optimal control is stated as follows.

Theorem 2.2 *Let $v(x, t)$ be the value function at initial time t,*

$$v(x, t) = \min_{U(\cdot) \in \mathscr{U}} \mathbb{E}\left[\int_t^T r(X(\tau), U(\tau), \tau) \, d\tau + g(X(T))\right],$$

(2.24)

and $v(x, T) = g(x)$ be the value function at terminal time T.

The dynamic programming principles states that, for any time s that satisfies $t < s \leq T$, the value function in state X at initial time t can be calculated backwards from $v(X_s, s)$,

$$v(x, t) = \min_{U(\cdot) \in \mathscr{U}} \mathbb{E}\left[\int_t^s r(X(\tau), U(\tau), \tau) \, d\tau + v(X_s, s)\right],$$

(2.25)

where $X_s = X(s)$.

In other words, DPP can be utilized to find the value function $v(x, t)$ for $t < s$ given the value function $v(X_s, s)$ at time s, without any information beyond time s (i.e., on time interval $[s, T]$).

2.2.2.5 Hamilton-Jacobi-Bellman Equation

A method to solve the value function $v(x, t)$ in (2.24) is to find a PDE whose solution is the value function itself. However, $v(x, t)$ may not be differentiable, and the PDE that can be solved by $v(x, t)$ does not exist. Nevertheless, the method in finding the HJB equation for deterministic optimal control may be applied here. The time horizon $[t, T]$ is divided into smaller subintervals of length δ. Then, the DPP in (2.25) is realized on each subinterval. This process leads to a PDE that is satisfied by $v(x, t)$.

The HJB equation is a PDE whose viscosity (i.e., weaker) solution is the value function $v(x, t)$. The derivation of the HJB equation starts with rewriting (2.25) by setting $s = t + \delta$,

$$v(x, t) = \min_{U(\cdot) \in \mathscr{U}} \mathbb{E}\left[\int_t^{t+\delta} r(X(\tau), U(\tau), \tau) \, d\tau + v(X(t + \delta), t + \delta) \right], \quad (2.26)$$

where $0 < \delta < T - t$.

Since v is a function of the stochastic process X, combining the equation for a solution of an SDE in (2.18) and the Ito stochastic differentiation rule in (2.19) yield

$$v(X(t + \delta), t + \delta) = v(x, t) + \int_t^{t+\delta} \left[\frac{\partial v}{\partial t}(X(\tau), \tau) + f \cdot \nabla_x v(X(\tau), \tau) \right] d\tau$$

$$+ \int_t^{t+\delta} \frac{1}{2} \text{Tr}[\sigma^\top \mathbb{H}_x(v(X(\tau), \tau))\sigma] \, d\tau$$

$$+ \int_t^{t+\delta} [\nabla_x v(X(\tau), \tau)]^\top \sigma \, dW(\tau).$$

Substituting this equation into (2.26) leads to

$$v(x, t) = \min_{U(\cdot) \in \mathscr{U}} \mathbb{E}\left[\int_t^{t+\delta} r(X(\tau), U(\tau), \tau) \, d\tau + v(x, t) \right.$$

$$+ \int_t^{t+\delta} \left[\frac{\partial v}{\partial t}(X(\tau), \tau) + f \cdot \nabla_x v(X(\tau), \tau) \right] d\tau$$

$$+ \int_t^{t+\delta} \frac{1}{2} \text{Tr}[\sigma^\top \mathbb{H}_x(v(X(\tau), \tau))\sigma] \, d\tau$$

$$\left. + \int_t^{t+\delta} [\nabla_x v(X(\tau), \tau)]^\top \sigma \, dW(\tau) \right].$$

Since the terms $v(x, t)$ and $\int_t^{t+\delta} \frac{\partial v}{\partial t}(X(\tau), \tau) \, d\tau$ do not depend on the control U, they can be taken outside of the min. Cancelling $v(x, t)$ on both sides and dividing the result by δ yield

$$\mathbb{E}\left[\frac{\int_t^{t+\delta} \frac{\partial v}{\partial t}(X(\tau), \tau) \, d\tau}{\delta}\right]$$

$$+ \min_{U(\cdot) \in \mathscr{U}} \mathbb{E}\left[\frac{\int_t^{t+\delta} r(X(\tau), U(\tau), \tau) \, d\tau}{\delta} + \frac{\int_t^{t+\delta} f \cdot \nabla_x v(X(\tau), \tau) \, d\tau}{\delta}\right.$$

$$\left.+ \frac{\int_t^{t+\delta} \frac{1}{2} \text{Tr}[\sigma^\top \mathbb{H}_x(v(X(\tau), \tau))\sigma] \, d\tau}{\delta} + \frac{\int_t^{t+\delta} [\nabla_x v(X(\tau), \tau)]^\top \sigma \, dW(\tau)}{\delta}\right] = 0.$$

Take the limit as $\delta \to 0$. For a continuous, nonnegative random variable ϕ, if $\mathbb{E}[\phi] = 0$, then $\phi = 0$. Also, for a continuous function ϕ,

$$\lim_{\delta \to 0} \frac{\int_t^{t+\delta} \phi(x(\tau), u(\tau), \tau) \, d\tau}{\delta} = \phi(x(t), u(t), t).$$

In addition, recognize that the expectation of the last term inside the min is zero. Consequently, the result is the HJB equation,

$$\frac{\partial v}{\partial t}(x, t) + \min_{u \in \mathbb{U}}\left[r(x, u, t) + f(x, u, t) \cdot \nabla_x v(x, t)\right.$$

$$\left.+ \frac{1}{2} \text{Tr}[\sigma^\top(x, u, t)\mathbb{H}_x(v(x, t))\sigma(x, u, t)]\right] = 0, \tag{2.27}$$

with terminal condition $v(x, T) = g(x)$, where $x = X(t)$ and $u = U(t)$.

The second term in (2.27) is known as the Hamiltonian H,

$$H(x, p, q, t) = \min_{u \in \mathbb{U}}\left[r(x, u, t) + f(x, u, t) \cdot p(t) + \frac{1}{2} \text{Tr}[\sigma^\top(x, u, t)q(t)\sigma(x, u, t)]\right], \tag{2.28}$$

where $p(t) = \nabla_x v(x, t)$ and $q(t) = \mathbb{H}_x[v(x, t)]$. When the optimal control u^* is realized, the Hamiltonian can be expressed as

$$\Phi(x, p, q, t) = r(x, u^*, t) + f(x, u^*, t) \cdot p(t) + \frac{1}{2} \text{Tr}[\sigma^\top(x, u^*, t)q(t)\sigma(x, u^*, t)]. \tag{2.29}$$

Hence, the HJB equation in (2.27) becomes

$$\frac{\partial v}{\partial t}(x, t) + \Phi(x, p, q, t) = 0. \tag{2.30}$$

If, for instance, $x(t) \in \mathbb{R}$, $f(x, u, t) \in \mathbb{R}$, and $\sigma \in \mathbb{R}$ is not a function of u, the HJB equation in (2.27) is reduced to

$$-\frac{\partial v}{\partial t}(x, t) - H(x, p, t) = \frac{\sigma^2}{2}\frac{\partial^2 v}{\partial x^2}(x, t), \tag{2.31}$$

with terminal condition $v(x, T) = g(x)$. The resulting Hamiltonian H becomes

$$H(x, p, t) = \min_{u \in \mathbb{U}} \left[r(x, u, t) + f(x, u, t) p(t) \right], \tag{2.32}$$

where $p(t) = \frac{\partial v}{\partial x}(x, t)$.

Suppose the value function $v(x, t)$ is twice-differentiable, and it solves (2.27) and (2.28). For any time $\tau > t$, the optimal control $U^*(\tau)$ is computed using the equation

$$U^*(\tau) = \arg\min_{u \in \mathbb{U}} \left[r(X^*(\tau), u, \tau) + f(X^*(\tau), u, \tau) \cdot \nabla_x v(X^*(\tau), \tau) \right.$$

$$\left. + \frac{1}{2} \mathrm{Tr}[\sigma^\top (X^*(\tau), u, \tau) \mathbb{H}_x(v(X^*(\tau), \tau)) \sigma(X^*(\tau), u, \tau)] \right], \tag{2.33}$$

and the corresponding optimal trajectory $X^*(\tau)$ solves the differential equation given by

$$dX^*(\tau) = f(X^*(\tau), U^*(\tau), \tau) \, d\tau + \sigma(X^*(\tau), U^*(\tau), \tau) \, dW(\tau), \tag{2.34}$$

with initial condition $X^*(t) = x$.

2.3 Differential Games

Differential games are dynamic games involving N players. These games are an extension of basic game theory to dynamic environments where game variables such as state and strategy vary with time. In a differential game, N players try to find the control function that optimizes their objective function subject to the state dynamics. In essence, each player in a differential game faces an optimal control problem in which the player also takes into account the control of other players in optimizing its objective function subject to its state dynamics.

In this section, two types of differential games are discussed: deterministic and stochastic differential games. In deterministic differential games, the state of the game is modeled as a deterministic process, whereas in stochastic differential games, the state of the game is modeled as a stochastic process. For more rigorous treatment of differential games, the reader may to refer to [6].

2.3.1   Deterministic Differential Games

Consider a set $\mathcal{N}$ of $N = |\mathcal{N}|$ players. The state of the system at time t is denoted as $x(t) : [0, T] \rightarrow \mathbb{R}^n$. The control of player i at time t is denoted as $u_i(t) : [0, T] \rightarrow \mathbb{R}^m$.

The dynamics of the state of the system is given by a system of n ordinary differential equations (ODEs)

$$dx(t) = f_i(x, u_i, u_{-i}, t) \, dt,$$
$$x(0) = x_0,$$
(2.35)

for $0 \le t \le T$ and $i \in \mathcal{N}$, where $x = x(t)$, $u_i = u_i(t)$, and $u_{-i} = u_{-i}(t)$ is the control profile containing the controls of other players besides player i, $u_{-i} = (u_1, \ldots, u_{i-1}, u_{i+1}, \ldots, u_N)$. The function $f_i(x, u_i, u_{-i}, t) : \mathbb{R}^n \times \mathbb{R}^m \times \mathbb{R}^m \times \cdots \times \mathbb{R}^m \times [0, T] \rightarrow \mathbb{R}^n$ is continuous with respect to all its arguments.

Each player i evaluates the quality of its control u_i is through the cost functional J_i

$$J_i(u_i, u_{-i}) = \int_0^T r_i(x, u_i, u_{-i}, t) \, dt + g_i(x(T)),$$
(2.36)

where $r_i(x, u_i, u_{-i}, t) : \mathbb{R}^n \times \mathbb{R}^m \times \mathbb{R}^m \times \cdots \times \mathbb{R}^m \times [0, T] \rightarrow \mathbb{R}$ refers to the running cost and $g_i(x) : \mathbb{R}^n \rightarrow \mathbb{R}$ refers to the terminal cost at the terminal state $x(T)$.

In a deterministic differential game, given the control profile u_{-i}, each player i seeks its optimal control u_i^* by solving the following deterministic optimal control problem,

$$J_i(u_i^*, u_{-i}) = \min_{u_i(\cdot) \in \mathcal{U}_i} J_i(u_i, u_{-i}) = \min_{u_i(\cdot) \in \mathcal{U}_i} \left[\int_0^T r_i(x, u_i, u_{-i}, t) \, dt + g_i(x(T)) \right],$$
(2.37)

subject to (2.35). That is, a deterministic differential game involves N players, and each player solves a deterministic optimal control problem of finding its optimal control given the controls of other players.

The value function $v_i(x, t)$ of player i associated with the solution of the deterministic optimal control problem faced by player i,

$$v_i(x, t) = \min_{u_i(\cdot) \in \mathcal{U}_i} J_i(u_i, u_{-i}) = \min_{u_i(\cdot) \in \mathcal{U}_i} \left[\int_0^T r_i(x, u_i, u_{-i}, t) \, dt + g_i(x(T)) \right],$$
(2.38)

is a solution of the HJB equation

$$\frac{\partial v_i}{\partial t}(x, t) + H_i(x, p_i, t) = 0,$$
(2.39)

with terminal condition $v_i(x, T) = g_i(x)$, where the Hamiltonian H_i is

$$H_i(x, p_i, t) = \min_{u_i \in U_i} \left[r_i(x, u_i, u_{-i}, t) + f_i(x, u_i, u_{-i}, t) \cdot p_i(t) \right], \tag{2.40}$$

and $p_i(t) = \nabla_x v_i(x, t)$.

The solution of a differential game, called the Nash equilibrium, refers to the control profile $u^* = (u_i^*, u_{-i}^*)$ that satisfies

$$J_i(u_i^*, u_{-i}^*) \leq J_i(u_i, u_{-i}^*), \tag{2.41}$$

for all $i \in \mathcal{N}$. Particularly, each player i seeks the optimal control u_i^* that solves the optimization problem

$$J_i(u_i^*, u_{-i}^*) = \min_{u_i(\cdot) \in \mathcal{U}_i} J_i(u_i, u_{-i}^*), \tag{2.42}$$

given that the other players play their optimal controls u_{-i}^*.

Let $v_i(x, t) = \min_{u_i(\cdot) \in \mathcal{U}_i} J_i(u_i, u_{-i}^*)$. Consequently, finding the Nash equilibrium of a differential game implies simultaneously solving N coupled HJB equations,

$$\frac{\partial v_i}{\partial t}(x, t) + H_i(x, p_i, t) = 0, \tag{2.43}$$

for all $i \in \mathcal{N}$, where the Hamiltonian H_i is

$$H_i(x, p_i, t) = \min_{u_i \in U_i} \left[r_i(x, u_i, u_{-i}^*, t) + f_i(x, u_i, u_{-i}^*, t) \cdot p_i(t) \right], \tag{2.44}$$

and $p_i(t) = \nabla_x v_i(x, t)$.

2.3.2 Stochastic Differential Games

The concepts from deterministic differential games can be extended to the case when the state of the system is a stochastic process.

Let the state of the system be a stochastic process $X(t) : [0, T] \times \Omega \to \mathbb{R}^n$. The control of player i at time t is denoted as $U_i(t) : [0, T] \to \mathbb{R}^m$. The state dynamics is given by a system of n stochastic differential equations (SDEs)

$$dX(t) = f_i(x, u_i, u_{-i}, t) \, dt + \sigma_i(x, u_i, u_{-i}, t) \, dW_i(t),$$
$$X(0) = x_0, \tag{2.45}$$

for $0 \leq t \leq T$ and $i \in \mathcal{N}$, where $x = X(t)$, $u_i = U_i(t)$, and $u_{-i} = U_{-i}(t)$ is the control profile containing the controls of other players besides player i, $u_{-i} =$

$(u_1, \ldots, u_{i-1}, u_{i+1}, \ldots, u_N)$. Meanwhile, $f_i(x, u_i, u_{-i}, t) : \mathbb{R}^n \times \mathbb{R}^m \times \mathbb{R}^m \times \cdots \times \mathbb{R}^m \times [0, T] \to \mathbb{R}^n$ and $\sigma_i(x, u_i, u_{-i}, t) : \mathbb{R}^n \times \mathbb{R}^m \times \mathbb{R}^m \times \cdots \times \mathbb{R}^m \times [0, T] \to \mathbb{R}^{n \times l}$ are continuous functions with respect to all their arguments, and $W_i(t) = (W_{i,1}(t), \ldots, W_{i,l}(t))^\top \in \mathbb{R}^l$ is a vector containing independent standard Wiener processes.

Let $r_i(x, u_i, u_{-i}, t) : \mathbb{R}^n \times \mathbb{R}^m \times \mathbb{R}^m \times \cdots \times \mathbb{R}^m \times [0, T] \to \mathbb{R}$ be the running cost and $g_i(x) : \mathbb{R}^n \to \mathbb{R}$ be the terminal cost at the terminal state $X(T)$. Each player i evaluates the performance of its control u_i through the cost functional J_i

$$J_i(u_i, u_{-i}) = \mathbb{E}\left[\int_0^T r_i(x, u_i, u_{-i}, t)\, dt + g_i(X(T))\right], \tag{2.46}$$

where $x = X(t)$, $u_i = U_i(t)$ refers to control of player i at time t, and $u_{-i} = U_{-i}(t)$ is the control profile at time t containing the controls of other players besides player i.

In a stochastic differential game, given the control profile u_{-i}, each player i aims to find its optimal control u_i^* from the set of admissible controls $\mathbb{U}_i$ by solving following stochastic optimal control problem,

$$J_i(u_i^*, u_{-i}) = \min_{u_i(\cdot) \in \mathbb{U}_i} J_i(u_i, u_{-i}) = \min_{u_i(\cdot) \in \mathbb{U}_i} \mathbb{E}\left[\int_0^T r_i(x, u_i, u_{-i}, t)\, dt + g_i(X(T))\right], \tag{2.47}$$

subject to (2.45). Specifically, a stochastic differential game involves N players, and each player solves a stochastic optimal control problem of finding its optimal control given the controls profile of other players.

Let the value function $v_i(x, t)$ associated with the solution of the stochastic optimal control problem be defined as

$$v_i(x, t) = \min_{u_i(\cdot) \in \mathbb{U}_i} J_i(u_i, u_{-i}) = \min_{u_i(\cdot) \in \mathbb{U}_i} \mathbb{E}\left[\int_0^T r_i(x, u_i, u_{-i}, t)\, dt + g_i(X(T))\right]. \tag{2.48}$$

Then, $v_i(x, t)$ is a solution of the HJB equation

$$\frac{\partial v_i}{\partial t}(x, t) + H_i(x, p_i, q_i, t) = 0, \tag{2.49}$$

with terminal condition $v_i(x, t) = g_i(x)$. Meanwhile, the Hamiltonian H_i is defined as

$$H_i(x, p_i, q_i, t) = \min_{u_i \in \mathbb{U}_i}\left[r_i(x, u_i, u_{-i}, t) + f_i(x, u_i, u_{-i}, t) \cdot p_i(t)\right.$$

$$\left. + \frac{1}{2}\mathrm{Tr}[\sigma_i^\top(x, u_i, u_{-i}, t) q_i(t) \sigma_i(x, u_i, u_{-i}, t)]\right], \tag{2.50}$$

where $p_i(t) = \nabla_x v_i(x, t)$ and $q_i(t) = \mathbb{H}_x[v_i(x, t)]$.

The Nash equilibrium or the solution of a differential game refers to the control profile $u^* = (u_i^*, u_{-i}^*)$ that satisfies

$$J_i(u_i^*, u_{-i}^*) \leq J_i(u_i, u_{-i}), \tag{2.51}$$

for all $i \in \mathcal{N}$. In a Nash equilibrium, each player i seeks the optimal control u_i^* that solves the optimization problem

$$J_i(u_i^*, u_{-i}^*) = \min_{u_i(\cdot) \in \mathbb{U}_i} J_i(u_i, u_{-i}^*), \tag{2.52}$$

given that the other players play their optimal controls u_{-i}^*.

Let $v_i(x, t) = \min_{u_i(\cdot) \in \mathbb{U}_i} J_i(u_i, u_{-i}^*)$. Therefore, the Nash equilibrium of a differential game can be found by simultaneously solving N coupled HJB equations,

$$\frac{\partial v_i}{\partial t}(x, t) + H_i(x, p_i, q_i, t) = 0, \tag{2.53}$$

for all $i \in \mathcal{N}$. The Hamiltonian H_i is now defined as

$$H_i(x, p_i, q_i, t) = \min_{u_i \in \mathbb{U}_i} \Big[r_i(x, u_i, u_{-i}^*, t) + f_i(x, u_i, u_{-i}^*, t) \cdot p_i(t)$$
$$+ \frac{1}{2} \text{Tr}[\sigma_i^\top(x, u_i, u_{-i}^*, t) q_i(t) \sigma_i(x, u_i, u_{-i}^*, t)] \Big], \tag{2.54}$$

where $p_i(t) = \nabla_x v_i(x, t)$ and $q_i(t) = \mathbb{H}_x[v_i(x, t)]$.

2.4 Mean Field Games

In this section, mean field games (MFGs) are formally examined. An MFG is a reformulated differential game where each player takes into account the mean field or the collective behavior of the players instead of the individual controls of every player in the game.

The discussion includes the motivation and mathematical theory of MFGs. In addition, analytic solution and numerical method in solving an MFG problem are both introduced. Furthermore, special forms of MFGs are presented: linear-quadratic MFG (LQMFG) and multiple-population MFG (MPMFG).

2.4.1 Background and Motivation

Consider a non-cooperative differential game with N players. Let $X(t) : [0, T] \rightarrow \mathbb{R}^n$ be the state of the game. The control of player i at time t is denoted as $u_i(t) :$

$[0, T] \rightarrow \mathbb{R}^m$, while control profile u_{-i} contains the control of players other than player i. Suppose that $r_i(x, u_i, u_{-i}, t) : \mathbb{R}^n \times \mathbb{R}^m \times \mathbb{R}^m \times \cdots \times \mathbb{R}^m \times [0, T] \rightarrow \mathbb{R}$ is a running cost function, $g_i(x) : \mathbb{R}^n \rightarrow \mathbb{R}$ is a terminal cost function or the cost function at terminal state $X(T)$, and $f_i(x, u_i, u_{-i}, t) : \mathbb{R}^n \times \mathbb{R}^m \times \mathbb{R}^m \times \cdots \times \mathbb{R}^m \times [0, T] \rightarrow \mathbb{R}^n$ is a drift function.

The goal of each player i is to find the control $u_i \in \mathscr{U}_i$ that minimizes its cost $J_i(u_i, u_{-i})$ subject to the state dynamics equation $dX(t)$,

$$
\begin{aligned}
\min_{u_i \in \mathscr{U}_i} \quad & J_i(u_i, u_{-i}) = \mathbb{E}\left[\int_0^T r_i(x, u_i, u_{-i}, t)\, dt + g_i(X(T)) \right], \\
\text{subject to} \quad & dX(t) = f_i(x, u_i, u_{-i}, t)\, dt + \sigma_i\, dW_i(t), \\
& X(0) = x_0,
\end{aligned}
\tag{2.55}
$$

where $W_i(t) \in \mathbb{R}^n$ denotes a standard Wiener process that accounts for the uncertainties in the state $x = X(t) \in \mathbb{R}^n$ and $\sigma_i > 0$ is a diffusion constant.

Any control u_i^* that solves the optimization problem in (2.55) is called an optimal control of player i. An optimal control u_i^* is a solution to (2.55) that minimizes the cost function subject to the state dynamics constraint. The formal definition of optimal control is stated as follows.

Definition 2.9 Let the value function $v_i(x, t)$ be defined mathematically as

$$
v_i(x, t) = \min_{u_i \in \mathscr{U}_i} J_i(u_i, u_{-i}),
$$

$\forall t \in [0, T]$. Then, the optimal control u_i^* satisfies

$$
J_i(u_i^*, u_{-i}) = \mathbb{E}\left[\int_0^T r_i(x, u_i^*, u_{-i}, t)\, dt + g_i(X(T)) \right] = v_i(x, t),
$$

$\forall t \in [0, T]$.

Meanwhile, if the value function $v_i(x, t)$ associated with the optimization problem in (2.55) is smooth enough (i.e., has continuous first-order derivative), then it is a viscosity solution to the following Hamilton-Jacobi-Bellman (HJB) equation [7],

$$
-\frac{\partial v_i}{\partial t}(x, t) - H_i(x, p_i, t) = \frac{\sigma_i^2}{2} \Delta_x v_i(x, t),
\tag{2.56}
$$

with terminal condition $v_i(x, T) = g_i(x)$. The term $\Delta_x v_i(x, t)$ denotes the Laplace of v_i with respect to x, which is the sum of the second derivatives of v_i with respect to every component of $x \in \mathbb{R}^n$. The term $H_i(x, p_i, t)$ is called the Hamiltonian, and it is defined mathematically as

$$H_i(x, p_i, t) = \min_{u_i \in \mathcal{U}_i} \left[r_i(x, u_i, u_{-i}, t) + f_i(x, u_i, u_{-i}, t) \cdot p(t) \right], \tag{2.57}$$

and $p_i = \nabla_x v_i(x, t)$.

When every player i faces the optimization problem in (2.55), and hence the corresponding HJB equation in (2.56), the resulting game is called a differential game [2]. Furthermore, when every player i solves its optimal control u_i^*, then it yields a solution to the game called Nash equilibrium. A Nash equilibrium is a solution concept to a game where no player can achieve a lesser cost by deviating from its optimal control given that other players maintain their optimal controls. A formal definition is given as follows.

Definition 2.10 Let J_i be the cost function of player i and (u_i^*, u_{-i}^*) be the control profile consisting of the control of player i and the controls u_{-i} of other players. Then, the control profile (u_i^*, u_{-i}^*) is a Nash equilibrium if and only if

$$J_i(u_i^*, u_{-i}^*) \leq J_i(u_i, u_{-i}^*), \forall u_i \in \mathcal{U}_i,$$

for every player i in the game.

The calculation of the Nash equilibrium of a differential game with N players involves solving N coupled HJB equations in (2.56). It becomes more complicated as the number of players increases because of increased interactions and coupling between the players. As a consequence, mean field games (MFGs) have been proposed to reformulate the game problem. MFGs were introduced by Lasry and Lions in [8] and have been applied in many applications in economics and engineering. MFGs can be utilized when the number of players is large and when the players are indistinguishable yet can have heterogeneous states. In an MFG, the aggregate effect of all the players is considered rather than the individual effect of each player.

Since the number of players in an MFG is large and the players are indistinguishable, the game can be seen as a representative or reference player playing against the mean field or aggregate behavior of other players. An MFG can be characterized by a pair of partial differential equations (PDEs): an HJB equation that corresponds to the evolution of the value function (i.e., optimized objective) of a player in response to the mean field, and a Fokker-Planck-Kolmogorov (FPK) equation that corresponds to the evolution of the mean field of players that are behaving optimally.

Consider a representative player with state $x \in \mathcal{X}$ and control $u \in \mathcal{U}$. Suppose m is the mean field or the distribution of the state of the players. An MFG can be expressed as a pair of HJB and FPK equations,

$$-\frac{\partial v}{\partial t}(x, t) - H(x, m, p, t) = \frac{\sigma^2}{2} \Delta_x v(x, t),$$

$$\frac{\partial m}{\partial t}(x, t) + \text{div}(f(x, u, m, t)m(x, t)) = \frac{\sigma^2}{2} \Delta_x m(x, t), \tag{2.58}$$

where Δ_x refers to the Laplace operator and div refers to the divergence operator. The function $H(x, m, p, t)$ is called the Hamiltonian, and it is defined mathematically as

$$H(x, m, p, t) = \min_{u \in \mathcal{U}} \left[r(x, u, m, t) + f(x, u, m, t) \cdot p(t) \right], \tag{2.59}$$

with boundary conditions $v(x, T) = g(x, m(x, T))$ and $m(x, 0) = m_0(x)$, where $p = \nabla_x v(x, t)$. The first equation in (2.58) is the HJB equation that characterizes the optimized reaction of a player with the mean field, while the second equation in (2.58) is the FPK equation that describes the evolution of the population that behaves optimally [9]. Note that the control profile u_{-i} found in differential games has now been replaced by mean field m in MFG to imply the dependence of the game on the collective behavior of the players rather than their individual influence.

The function $m(x, t)$ refers to the mean field, and it corresponds to the distribution of the states of the players with respect to time. The formal definition of $m(x, t)$ is given as follows.

Definition 2.11 The mean field $m(x, t)$ denotes the probability distribution of the players with state x at time t. Mathematically,

$$m(x, t) = \lim_{N \to \infty} \frac{1}{N} \left(\sum_{i=1}^{N} \delta_{x_i = x} \right), \tag{2.60}$$

where $\delta = 1$ if $x_i = x$ and $\delta = 0$ if $x_i \neq x$.

The following theorem defines an MFG problem and provides insightful proof of the HJB and FPK equations. For more comprehensive analysis regarding MFGs, the reader may refer to [8].

Theorem 2.3 *Consider a non-cooperative game among large number of indistinguishable players. If every player faces the optimization problem*

$$\min_{u \in \mathcal{U}} \quad J(u, m) = \mathbb{E} \left[\int_0^T r(x, u, m, t) \, dt + g(X(T), m(x, T)) \right],$$
$$subject \ to \quad dX(t) = f(x, u, m, t) \, dt + \sigma \, dW(t), \tag{2.61}$$
$$X(0) = x_0,$$

the equivalent non-cooperative mean field game is represented by the pair of HJB and FPK equations,

$$-\frac{\partial v}{\partial t}(x, t) - H(x, m, p, t) = \frac{\sigma^2}{2} \Delta_x v(x, t),$$
$$\frac{\partial m}{\partial t}(x, t) + div \left[\frac{\partial H}{\partial p}(x, m, p, t) m(x, t) \right] = \frac{\sigma^2}{2} \Delta_x m(x, t), \tag{2.62}$$

where the boundary conditions are $v(x, T) = g(x, m(x, T))$ and $m(x, 0) = m_0(x)$, and the Hamiltonian $H(x, m, p, t) = \min_{u \in \mathcal{U}} [r(x, u, m, t) + f(x, u, m, t) \cdot p(t)]$, where $p(t) = \nabla_x v(x, t)$.

In addition, the optimal control $u^*(x, t)$ is the solution to the equation

$$f(x, u^*, m, t) = \frac{\partial H}{\partial p}(x, m, p, t), \tag{2.63}$$

such that the mean field m when $u = u^*$ in (2.61) coincides with the solution m^* of the FPK equation in (2.62).

Proof The following derivation of the HJB equation follows the derivation of HJB equation in stochastic optimal control discussed in Sect. 2.2.

Let the value function $v(x, t) = \min_{u \in \mathcal{U}} J(u, m)$. Applying the dynamic programming principle (DPP) for stochastic optimal control to $v(x, t)$ leads to

$$v(x, t) = \min_{u \in \mathcal{U}} \mathbb{E} \left[\int_t^{t+\delta} r(x, u, m, \tau) \, d\tau + v(X(t + \delta), t + \delta) \right],$$

where $0 < \delta < T$.

Using Ito stochastic differentiation rule, the term $v(X(t + \delta), t + \delta)$ can be written as

$$v(X(t + \delta), t + \delta) = v(x, t) + \int_t^{t+\delta} \left[\frac{\partial v}{\partial t}(X(\tau), \tau) + f(x, u, m, \tau) \cdot \nabla_x v(X(\tau), \tau) \right] d\tau$$

$$+ \int_t^{t+\delta} \frac{\sigma^2}{2} \Delta_x v(X(\tau), \tau) \, d\tau$$

$$+ \int_t^{t+\delta} \sigma \nabla_x v(X(\tau), \tau) \, dW(\tau).$$

Substituting this equation into the first equation of $v(x, t)$, taking the terms $v(x, t)$ and $\int_t^{t+\delta} \frac{\partial v}{\partial t}(X(\tau), \tau) \, d\tau$ outside of the min since they are independent of u, and cancelling $v(x, t)$ on both sides result in

$$\int_t^{t+\delta} \frac{\partial v}{\partial t}(X(\tau), \tau) \, d\tau$$

$$+ \min_{u \in \mathcal{U}} \mathbb{E} \left[\int_t^{t+\delta} [r(x, u, m, \tau) \, d\tau + f(x, u, m, \tau) \cdot \nabla_x v(X(\tau), \tau)] \, d\tau \right.$$

$$\left. + \int_t^{t+\delta} \frac{\sigma^2}{2} \Delta_x v(X(\tau), \tau) \, d\tau + \int_t^{t+\delta} \sigma \nabla_x v(X(\tau), \tau) \, dW(\tau) \right] = 0.$$

Dividing the result by δ, getting the limit as $\delta \rightarrow 0$, and evaluating the expectation lead to the HJB equation,

$$-\frac{\partial v}{\partial t}(x, t) - H(x, m, p, t) = \frac{\sigma^2}{2} \Delta_x v(x, t),$$

with $H(x, m, p, t) = \min_{u \in \mathscr{U}} \left[r(x, u, m, t) + f(x, u, m, t) \cdot p(t) \right]$ and $p(t) = \nabla_x v(x, t)$. This concludes the derivation of the HJB equation in (2.62). If $u = u^*$, the Hamiltonian can be written as

$$H(x, m, p, t) = r(x, u^*, m, t) + f(x, u^*, m, t) \cdot p(t).$$

Consequently,

$$f(x, u^*, m, t) = \frac{\partial H}{\partial p}(x, m, p, t).$$

Meanwhile, the derivation of the FPK equation starts with letting ϕ be a twice-differentiable function of a stochastic process $X(t) = x$ with state dynamics $dX(t)$ governed by (2.61).

Applying Ito differentiation rule to get $d\phi(x)$ yields

$$d\phi(x) = \left[f(x, u, m, t) \cdot \nabla_x \phi(x) + \frac{\sigma^2}{2} \Delta_x \phi(x) \right] dt + \nabla_x \phi(x) \sigma \, dW(t).$$

Dividing by dt and taking the expectation give

$$\mathbb{E}\left[\frac{d\phi(x)}{dt} \right] = \mathbb{E}\left[f(x, u, m, t) \cdot \nabla_x \phi(x) + \frac{\sigma^2}{2} \Delta_x \phi(x) \right],$$

since $\mathbb{E}[\nabla_x \phi(x) \sigma \, dW(t)] = 0$.

Rewriting in terms of the mean field $m(x, t)$ and using integration by parts lead to

$$\int_{\mathscr{X}} \frac{\partial \phi}{\partial x}(t) m(x, t) \, dx = \int_{\mathscr{X}} [f(x, u, m, t) \cdot \nabla_x \phi(x)] m(x, t) \, dx$$
$$+ \int_{\mathscr{X}} \frac{\sigma^2}{2} \Delta_x \phi(x) m(x, t) \, dx,$$

$$\int_{\mathscr{X}} \frac{\partial m}{\partial x}(x, t) \phi(x) \, dx = \int_{\mathscr{X}} -\mathrm{div}\big(f(x, u, m, t) m(x, t)\big) \phi(x, t) \, dx$$
$$+ \frac{\sigma^2}{2} \int_{\mathscr{X}} \Delta_x m(x, t) \phi(x) \, dx.$$

Differentiating the last equation with respect to x, dividing out $\phi(x)$, and assuming $u = u^*$ yield the FPK equation in (2.62),

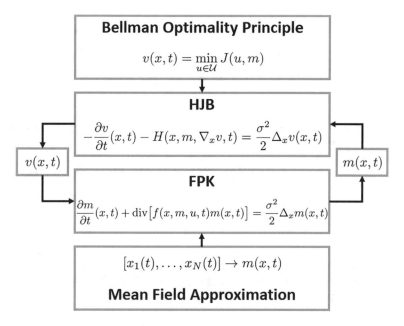

Fig. 2.2 Mechanics of a mean field game

$$\frac{\partial m}{\partial t}(x, t) + \operatorname{div}\left[\frac{\partial H}{\partial p}(x, m, p, t)m(x, t)\right] = \frac{\sigma^2}{2}\Delta_x m(x, t).$$

Figure 2.2 illustrates the mechanics of an MFG. According to the Bellman principle of optimality, "[an] optimal policy has the property that whatever the initial state and initial decision are, the remaining decisions must constitute an optimal policy with regard to the state resulting from the first decision" [5]. The dynamic programming principle (DPP) takes advantage of this principle by providing a method of solving complex problems by breaking them down into smaller subproblems. DPP transforms an optimization problem into an HJB equation, as shown in the proof of Theorem 2.3. Meanwhile, the evolution of the distribution of the states is derived by getting the mean field approximation of the state dynamic equation that results to an FPK equation, as shown in the proof of Theorem 2.3. Together, the HJB and FPK equations make up an MGF system in which the solution of one partial differential equation is required to solve to other partial differential equation and vice versa.

2.4.2 Analytic Solution

The adjoint method is a method of solving optimization problems with partial differential equation (PDE) constraints [10, 11]. The analytic method starts with introducing an adjoint variable, $v(x, t)$, which also corresponds to the value function

$$v(x, t) = \min_{u \in \mathscr{U}} J(u, m). \tag{2.64}$$

Using this adjoint variable, the FPK equation is appended to the original cost function $J(u, m)$. Thus, the resulting MFG problem optimizes the extended cost function,

$$
\begin{aligned}
\mathscr{J}(u, m, v) = {} & \int_0^T \int_{\mathscr{X}} r(x, u, m, t) m(x, t) dx\, dt + \int_{\mathscr{X}} g(X(T), m(x, T)) m(x, T)\, dx \\
& + \int_0^T \int_{\mathscr{X}} v(x, t) \Big(-\frac{\partial m}{\partial t}(x, t) - \mathrm{div}\big(f(x, u, m, t) m(x, t) \big) \\
& + \frac{\sigma^2}{2} \Delta_x m(x, t) \Big)\, dx\, dt.
\end{aligned}
\tag{2.65}
$$

There exists a pair (m, u) that minimizes the extended cost function in (2.65) if there is a v such that (m, u, v) is a stationary solution. Thus, the following optimality conditions are necessary:

$$\frac{\partial \mathscr{J}}{\partial u} = 0, \quad \frac{\partial \mathscr{J}}{\partial m} = 0, \quad \text{and} \quad \frac{\partial \mathscr{J}}{\partial v} = 0. \tag{2.66}$$

Now, consider the optimal control problem

$$\min_{m, u} J(u, m), \tag{2.67}$$

subject to

$$\frac{\partial m}{\partial t}(x, t) + \mathrm{div}\big(f(x, u, m, t) m(x, t) \big) = \frac{\sigma^2}{2} \Delta_x m(x, t). \tag{2.68}$$

Solving $\frac{\partial \mathscr{J}}{\partial u} = 0$ yields

$$\frac{\partial r}{\partial u}(x, u, m, t) + p(t) \frac{\partial f}{\partial u}(x, u, m, t) = 0, \tag{2.69}$$

where $p(t) = \nabla_x v(x, t)$.

Meanwhile, $\frac{\partial \mathscr{J}}{\partial m} = 0$ results to

$$-\frac{\partial v}{\partial t}(x, t) - H(x, m, p, t) = \frac{\sigma^2}{2} \Delta_x v(x, t). \tag{2.70}$$

Finally, $\frac{\partial \mathscr{J}}{\partial v} = 0$ yields

$$\frac{\partial m}{\partial t}(x, t) + \text{div}\big(f(x, u, m, t)m(x, t)\big) = \frac{\sigma^2}{2}\Delta_x m(x, t).\tag{2.71}$$

Hence, (2.69), (2.70), and (2.71) are solved simultaneously to find the stationary solution (m, u, v).

2.4.3 Numerical Methods

The numerical method utilized in this work to implement the computation of the solution in the previous subsection is based on the numerous work in numerical MFG, such as [11, 12], and [13].

Consider a bounded region $[0, X_{\max}] \times [0, T_{\max}]$ over which independent variables x and t of the PDEs are defined. In order to implement a numerical method to solve the PDEs, the region is converted into a finite grid of points. Given positive integers L and M, a space step is defined as $\Delta x = \frac{X_{\max}}{L}$, and a time step is defined as $\Delta t = \frac{T_{\max}}{M}$. Hence, the grid of points are defined by

$$x_j = j\Delta x, \, j = 0, \ldots, L,$$

and
$$t_k = k\Delta t, \, k = 0, \ldots, M.$$

Furthermore, for any function ρ defined over the space-time region, $\rho_j^k = \rho(x_j, t_k)$.

The finite difference (FD) method is a numerical method to approximate and solve PDEs. This method uses finite differences to approximate derivatives. The following operators denote the first-order derivatives in x and t.

$$(D_t\rho)^k = \frac{\partial\rho}{\partial t} = \frac{\rho^{k+1} - \rho^k}{\Delta t},$$

and
$$(D_x\rho)_j = \frac{\partial\rho}{\partial x} = \frac{\rho_{j+1} - \rho_j}{\Delta x}.$$

The next operator can be used to denote the second-order derivative in x,

$$(\Delta_x\rho)_j = \frac{\partial^2\rho}{\partial x^2} = \frac{\rho_{j+1} - 2\rho_j + \rho_{j-1}}{(\Delta x)^2}.$$

Applying these FD operators to the MFG equations in (2.62) and setting $\varepsilon = \sigma^2/2$ produce the following discrete MFG equations,

$$-(D_t v_j)^k - \bar{H}(x_j, m_j^{k+1}, (D_x v^k)_j) = \varepsilon(\Delta_x v^k)_j,$$
$$(D_t m_j)^k + \mathcal{T}_j(v^k, m^{k+1}) = \varepsilon(\Delta_x m^{k+1})_j,\tag{2.72}$$

with boundary conditions $v_0^k = v_L^k$, $v_j^M = g(m_j^M)$, $m_0^k = m_L^k$, and $m_j^0 = \bar{m}_0(x_j)$. The term $\bar{H}$ is the discrete Hamiltonian and $\mathcal{T}_j$ is the discrete transport operator.

2.4.4   Linear-Quadratic Mean Field Games

Linear-quadratic mean field games (LQMFGs) are a subclass of MFGs in which the cost functional is quadratic in all state variables, control variables, and mean field terms [14].

Let $(\Omega, \mathscr{F}, P)$ be a complete probability space. The initial states of the players $x_0 \in \mathbb{R}^n$ are independent, identically distributed random vectors.

The main goal of each player is to find the optimal control $u^*(t)$ by minimizing its cost functional through controlling its state dynamics. The cost functional of a player is given by

$$J(u, m) = \mathbb{E}\left[\frac{1}{2}\int_0^T \left[x^\top(t)Q_t x(t) + u^\top(t)R_t u(t)\right]dt + \frac{1}{2}x^\top(T)Q_T x^\top(T)\right]$$

$$+ \mathbb{E}\left[\frac{1}{2}\int_0^T \left(x(t) - S_t m(t)\right)^\top \bar{Q}_t\left(x(t) - S_t m(t)\right)dt\right]$$

$$+ \mathbb{E}\left[\frac{1}{2}\left(x(T) - S_T m(T)\right)^\top \bar{Q}_T\left(x(T) - S_T m(T)\right)\right],$$

(2.73)

where Q, $\bar{Q}$, R, and S are bounded deterministic matrix-valued functions of time, and $m(t) = \mathbb{E}[x^*(t)]$ with $x^*(t)$ corresponding to the trajectory at optimal control $u^*(t)$. Meanwhile, the state dynamics of a player is modeled by the stochastic differential equation

$$dx(t) = \left[A(t)x(t) + B(t)u(t) + \bar{A}(t)m(t)\right]dt + \sigma(t)\,dW(t),$$

$$x(0) = x_0,$$

(2.74)

where A, B, and $\bar{A}$ are bounded deterministic matrix-valued functions of time, and σ is a function of time.

For the LQMFG, the optimal control $u^*(t)$ is given by

$$u^*(t) = -R_t^{-1}B_t^\top p(t),$$

(2.75)

where (x^*, p) solves the following system of differential equations:

$$dx^*(t) = [A_t x^*(t) - B_t R_t^{-1}B_t^\top p(t) + \bar{A}_t m(t)]dt + \sigma(t)\,dW(t),$$

$$x^*(0) = x_0,$$

$$-\frac{d\omega(t)}{dt} = A_t^\top \omega(t) + (Q_t + \bar{Q}_t)x^*(t) - \bar{Q}_t S_t m(t),$$

$$\omega(T) = (Q_T + \bar{Q}_T)x^*(T) - \bar{Q}_t S_t m(t),$$

(2.76)

such that $p(t) = \mathbb{E}[\omega(t)|\mathcal{F}(t)]$. Meanwhile, $(m, \mathbb{E}[p(t)]) = (\xi, \eta)$ solves the following system of differential equations:

$$\frac{d\xi(t)}{dt} = (A_t + \bar{A}_t)\xi(t) - B_t R_t^{-1} B_t^\top \eta(t),$$

$$\xi(0) = \mathbb{E}[x_0],$$

$$-\frac{d\eta(t)}{dt} = [Q_t + \bar{Q}_t(I - S_t)]\xi(t) + A_t^\top \eta(t),$$

$$\eta(T) = (Q_T + \bar{Q}_T(I - S_T))\eta(T),$$

(2.77)

where I is the identity matrix.

2.4.5 Multiple-Population Mean Field Games

In this subsection, the multiple-population MFG (MPMFG) framework presented in [15] is introduced. It is an extension of the single-population MFG to multiple-population games in which each population tries to optimize their objective function.

Consider a non-cooperative game among a large number of players that can be divided into P populations. Let the state $x \in \mathcal{X}$. Denote mean field vector $m = (m_i)_{i=1,\dots,P}^\top$ and control vector $u = (u_i)_{i=1,\dots,P}^\top$. Assume that each population considers that the distribution of all populations are fixed or predictable and tries to minimize its own individual cost. The goal of a player in population i is to minimize the population cost $J_i(u_i, m)$ subject to the population state dynamics $dx(t)$.

In an MFG problem with P populations, population i faces the problem of finding a pair (u_i^*, m_i^*) that solves

$$\min_{u_i \in \mathcal{U}_i} J_i(u_i, m) = \mathbb{E}\left[\int_0^T r_i(x, u_i, m)\, dt + g_i(x, m_T)\right], \quad (2.78)$$

subject to the mean field dynamics

$$\frac{\partial m_i}{\partial t}(x, t) + \mathrm{div}\big(f_i(x, u_i, m)m_i(x, t)\big) = \frac{\sigma_i^2}{2}\Delta_x m_i(x, t),$$

$$\frac{\partial m_p}{\partial t}(x, t) + \mathrm{div}\big(f_p(x, u_p^*, m)m_p(x, t)\big) = \frac{\sigma_p^2}{2}\Delta_x m_p(x, t),$$

(2.79)

for $p \neq i$, with boundary conditions $m_i(x, 0) = m_{i,0}$ and $m(x, T) = m_T$ for all P populations.

The pair (u_i^*, m_i^*) that solves the MFG problem above is also a solution of the HJB and FPK equations for population i. Hence, the mean field m_i of the MFG problem must coincide with the mean field m_i^* solution of the FPK equation for population i.

2.5   Mean-Field-Type Games

The concept of mean-field-type games (MFTGs) is studied in this section. An MFTG is relaxed version of MFG in which the number of players is not necessarily large and the players are not necessarily indistinguishable. While there many forms of MFTG, this section focuses on linear-quadratic forms in which the state dynamic equation is linear in the system variables and the cost functional is quadratic in the system variables. The discussion includes linear-quadratic mean-field-type control (LQMFTC), which is a one-player optimal control problem, and linear-quadratic MFTG (LQMFTG), which is a many-player MFTG problem.

2.5.1   Background

In MFGs, the following assumptions are usually made [16]: the number of players is infinitely high; the players are indistinguishable; and a player does not affect the mean field term. However, in engineering applications, these assumptions are difficult to prove. Usually, the number of players or decision makers in engineering applications is finite. Also, the players are not necessarily anonymous. Finally, a player can have a significant effect on the mean field term. A methodology that relaxes the assumptions made in an MFG is called a mean-field-type game (MFTG). An MFTG model has an instantaneous cost and/or the state dynamic equation that involves not only the mean field associated with the distribution of the states but also the joint distributions of the state-action pairs.

Consider an MFTG with $N \geq 2$ players. An MFTG model has players with individual mean field terms defined by the probability distributions $m_i(t) = P(x_i)$ and $\mu_i(t) = P(x_i, u_i)$, and population mean field terms $m_{-i}(t) = \frac{\sum_{k \neq i} \delta_{x_i}}{N-1}$ and $\mu_{-i}(t) = \frac{\sum_{k \neq i} \delta_{(x_i, u_i)}}{N-1}$. The term m_i is a probability distribution of state x_i while μ_i is a probability distribution of state-action pair (x_i, u_i). When the number of interacting players is finite, the population mean field terms characterize the effect of other players. In this MFTG model, the influence of a player on its individual mean field terms is non-negligible, regardless of the large number of interacting players.

An MFTG problem involves optimizing the cost of player i,

$$J_i(u_i) = \mathbb{E}\left[\int_0^T r_i(x_i, u_i, \mu_i, \mu_{-i}) \, dt + g_i(x_i(T), m_i(T), m_{-i}(T)) \right], \quad (2.80)$$

subject to player i state dynamics,

$$dx_i(t) = f_i(x_i, u_i, \mu_i, \mu_{-i}) \, dt + \sigma_i(x_i, u_i, \mu_i, \mu_{-i}) \, dW_i(t),$$
$$x_i(0) \sim m_i(0),$$

(2.81)

over $[0, T]$, $T > 0$, where r_i, g_i, f_i, and σ_i are measurable functions and $W_i(t)$ is a standard Wiener process.

2.5.2 Linear-Quadratic Mean-Field-Type Control

The integration of variance in MFTGs allows for modeling of risk-aware controls that minimize the risk or the associated variance of the state [17]. Consider a one-player game in which the player seeks the optimal control that minimizes the risk or variance of the system. The player faces the following stochastic optimal control problem,

$$\min_{u \in \mathcal{U}} \mathbb{E} \left[\frac{1}{2} \int_0^T \left[q(t)x^2(t) + r(t)u^2(t) + \bar{q}(t)\mathbb{E}[x(t)]^2 + \bar{r}(t)\mathbb{E}[u(t)]^2 \right] dt \right.$$
$$\left. + \frac{1}{2}q(T)x^2(T) + \frac{1}{2}\bar{q}(T)\mathbb{E}[x(T)]^2 \right],$$

(2.82)

subject to

$$dx(t) = \left[b_0(t) + b_1(t)x(t) + \bar{b}_1(t)\mathbb{E}[x(t)] + b_2(t)u(t) + \bar{b}_2(t)\mathbb{E}[u(t)] \right] dt$$
$$+ \sigma_0(t) \, dW(t),$$
$$x(0) = x_0,$$

(2.83)

where $q(t), q(t) + \bar{q}(t) \in \mathbb{R}_{\geq 0}$, $r(t), r(t) + \bar{r}(t) \in \mathbb{R}_{\geq 0}$, for all $t \in [0, T]$, $b_0(t)$, $b_1(t)$, $\bar{b}_1(t)$, $b_2(t)$, $\bar{b}_2(t)$, $\sigma_0(t) : \mathbb{R} \to \mathbb{R}$ are scalar functions of time, and $W(t)$ refers to a standard Wiener process. This one-player game in which the player optimizes the cost functional subject to the state dynamic equation is known as a linear-quadratic mean-field-type control (LQMFTC).

The optimal control law and optimal cost of the LQMFTC problem are given by

$$u^*(t) = \mathbb{E}[u^*(t)] - \frac{b_2(t)}{r(t)}\alpha(t)(x(t) - \mathbb{E}[x(t)]),$$

$$\mathbb{E}[u^*(t)] = -\frac{b_2(t) + \bar{b}_2(t)}{r(t) + \bar{r}(t)}[\beta(t)\mathbb{E}[x(t)] + \gamma(t)],$$

(2.84)

$$L(x_0, u^*) = \frac{1}{2}\alpha(0)(x_0 - \mathbb{E}[x_0])^2 + \frac{1}{2}\beta(0)\mathbb{E}[x_0]^2 + \gamma(0)\mathbb{E}[x_0] + \delta(0),$$

where $\alpha(t)$, $\beta(t)$, $\gamma(t)$, and $\delta(t)$ solve the following differential equations:

$$\dot{\alpha}(t) = -2b_1(t)\alpha(t) - q(t) + \frac{b_2^2(t)}{r(t)}\alpha^2(t),$$

$$\dot{\beta}(t) = -2(b_1(t) + \bar{b}_1(t))\beta(t) - q(t) - \bar{q}(t) + \frac{(b_2(t) + \bar{b}_2(t))^2}{r(t) + \bar{r}(t)}\beta^2(t),$$

$$\dot{\gamma}(t) = -b_0\beta(t) - (b_1(t) - \bar{b}_1(t))\gamma(t) + \frac{(b_2(t) + \bar{b}_2(t))^2}{r(t) + \bar{r}(t)}\beta(t)\gamma(t),$$

$$\dot{\delta}(t) = -b_0(t)\gamma(t) - \frac{\alpha(t)}{2}\sigma_0^2(t) + \frac{(b_2(t) + \bar{b}_2(t))^2}{2(r(t) + \bar{r}(t))}\gamma^2(t),$$

(2.85)

where $\alpha(T) = q(T)$, $\beta(T) = q(T) + \bar{q}(T)$, and $\gamma(T) = \delta(T) = 0$, whenever these equations admit a solution with positive α and β.

2.5.3   Linear-Quadratic Mean-Field-Type Games

The concept of LQMFTC can be extended to an N-player game in which each player seeks to minimize its cost by finding the optimal control subject to the state dynamics equation.

In a non-cooperative linear-quadratic MFTG (LQMFTG), each player solves the following LQMFTG problem,

$$\min_{u_i \in \mathscr{U}_i} \mathbb{E}\left[\frac{1}{2}\int_0^T \left[q_i(t)x^2(t) + r_i(t)u_i^2(t) + \bar{q}_i(t)\mathbb{E}[x(t)]^2 + \bar{r}_i^2(t)\mathbb{E}[u_i(t)]^2\right] dt\right.$$

$$\left. + \frac{1}{2}q_i(T)x^2(T) + \frac{1}{2}\bar{q}_i(T)\mathbb{E}[x(T)]^2\right],$$

(2.86)

subject to

$$dx(t) = \left[b_0(t) + b_1(t)x(t) + \bar{b}_1(t)\mathbb{E}[x(t)] + \sum_{j=1}^N b_{2j}(t)u_j(t) + \sum_{j=1}^N \bar{b}_{2j}(t)\mathbb{E}[u_j(t)]\right] dt$$

$$+ \sigma_0(t)\,dW(t),$$

$$x(0) = x_0,$$

(2.87)

for all $i = 1, \ldots, N$, where $q_i(t)$, $q_i(t) + \bar{q}_i(t) \in \mathbb{R}_{\geq 0}$, $r_i(t)$, $r_i(t) + \bar{r}_i(t) \in \mathbb{R}_{\geq 0}$, for all $t \in [0, T]$, and $b_0(t)$, $b_1(t)$, $\bar{b}_1(t)$, $b_{2j}(t)$, $\bar{b}_{2j}(t)$, $\sigma_0(t) \in \mathbb{R}$, for $j = 1, \ldots, N$, and $W(t)$ refers to a standard Wiener process.

The authors of [17] proposed a method to find the solution to an LQMFTG problem. Applying this method to the LQMFTG in (2.86)–(2.87), the optimal control law and optimal cost are given by

$$u_i^*(t) = \mathbb{E}[u_i^*(t)] - \frac{b_{2i}(t)}{r_i(t)}\alpha_i(t)(x(t) - \mathbb{E}[x(t)]),$$

$$\mathbb{E}[u_i^*(t)] = -\frac{b_{2i}(t) + \bar{b}_{2i}(t)}{r_i(t) + \bar{r}_i(t)}[\beta_i(t)\mathbb{E}[x(t)] + \gamma_i(t)],$$

$$L(x_0, u_i^*, u_{-i}^*) = \frac{1}{2}\alpha_i(0)(x_0 - \mathbb{E}[x_0])^2 + \frac{1}{2}\beta_i(0)\mathbb{E}[x_0]^2 + \gamma_i(0)\mathbb{E}[x_0] + \delta_i(0),$$

$$(2.88)$$

for all $i = 1, \ldots, N$, where $\alpha_i(t)$, $\beta_i(t)$, $\gamma_i(t)$, and $\delta_i(t)$ solve the following differential equations:

$$\dot{\alpha}_i(t) = -2b_1(t)\alpha_i(t) - q_i(t) + 2\alpha_i(t)\sum_{j=1\neq i}^{N}\frac{b_{2j}^2(t)}{r_j(t)}\alpha_j^2(t) + \frac{b_{2i}^2(t)}{r_i(t)}\alpha_i^2(t),$$

$$\dot{\beta}_i(t) = -2[b_1(t) + \bar{b}_1(t)]\beta_i(t) - q_i(t) - \bar{q}_i(t) + 2\beta_i(t)\sum_{j=1\neq i}^{N}\frac{(b_{2j}(t) + \bar{b}_{2j}(t))^2}{r_j(t) + \bar{r}_j(t)}\beta_j^2(t)$$

$$+ \frac{(b_{2i}(t) + \bar{b}_{2i}(t))^2}{r_i(t) + \bar{r}_i(t)}\beta_i^2(t),$$

$$\dot{\gamma}_i(t) = -b_0\beta_i(t) - [b_1(t) - \bar{b}_1(t)]\gamma_i(t) + \beta_i(t)\sum_{j=1\neq i}^{N}\frac{(b_{2j}(t) + \bar{b}_{2j}(t))^2}{r_j(t) + \bar{r}_j(t)}\gamma_j(t)$$

$$+ \gamma_i(t)\sum_{j=1\neq i}^{N}\frac{(b_{2j}(t) + \bar{b}_{2j}(t))^2}{r_j(t) + \bar{r}_j(t)}\beta_j(t) + \frac{(b_{2i}(t) + \bar{b}_{2i}(t))^2}{r_i(t) + \bar{r}_i(t)}\beta_i(t)\gamma_i(t),$$

$$\dot{\delta}_i(t) = -b_0(t)\gamma_i(t) - \frac{1}{2}\alpha_i(t)\sigma_0^2(t) + \gamma_i(t)\sum_{j=1\neq i}^{N}\frac{(b_{2j}(t) + \bar{b}_{2j}(t))^2}{r_j(t) + \bar{r}_j(t)}\gamma_j(t)$$

$$+ \frac{1}{2}\frac{(b_{2i}(t) + \bar{b}_{2i}(t))^2}{r_i(t) + \bar{r}_i(t)}\gamma_i(t)^2,$$

$$(2.89)$$

where $\alpha_i(T) = q_i$, $\beta_i(T) = q_i + \bar{q}_i$, and $\gamma_i(T) = \delta_i(T) = 0$, whenever these equations admit a solution with positive α_i and β_i.

References

1. M. Maschler, E. Solan, S. Zamir, *Game Theory* (Cambridge University Press, Cambridge, 2013)
2. Z. Han, D. Niyato, W. Saad, T. Basar, A. Hjørungnes, *Game Theory in Wireless and Communication Networks: Theory, Models and Applications* (Cambridge University Press, Cambridge, 2012)
3. W.H. Fleming, R.W. Rishel, *Deterministic and Stochastic Optimal Control* (Springer, Berlin, 1975)
4. G.E. Kolosov, *Optimal Design of Control Systems: Stochastic and Deterministic Problems* (Dekker, New York, 1999)
5. R. Bellman. *Dynamic Programming* (Princeton University Press, Princeton, 1957)
6. A. Friedman, *Differential Games* (Dover Publications, Mineola, 2006)
7. M. Bardi, I. Capuzzo-Dolcetta, *Optimal Control and Viscosity Solutions of Hamilton-Jacobi-Bellman Equations* (Springer, Berlin, 2008)
8. J.M. Lasry, P.L. Lions, Mean field games. Jpn J. Math. **2**(1), 229–260 (2007)
9. D.A. Gomes, L. Nurbekyan, E.A. Pimentel, Economic models and mean-field game theory, *30° Colóquio Brasileiro de Matemática*, Rio de Janeiro, Brazil (2015)
10. Y. Cao, S. Li, L. Petzold, R. Serban, Adjoint sensitivity analysis for differential-algebraic equations: the adjoint DAE system and its numerical solution. SIAM J. Sci. Comput. **23**(3), 1076–1089 (2003)
11. J.M. Schulte, *Adjoint Methods for Hamilton-Jacobi-Bellman Equations*, Diploma Thesis, University of Munster, Germany (2010)
12. Y. Achdou, Finite difference methods for mean field games, in *Hamilton-Jacobi Equations: Approximations, Numerical Analysis and Applications* (Springer, Berlin, 2013)
13. Y. Achdou, M. Lauriere, Mean field game and applications: numerical aspects, in *Mean Field Games* (Springer International Publishing, New York, 2020)
14. A. Bensoussan, K.C.J. Sung, S.C.P. Yam, S.P. Yung, Linear-quadratic mean field games. J. Optim. Theory Appl. **169**(2), 496–529 (2016)
15. A. Bensoussan, T. Huang, M. Lauriere, Mean field control and mean field game models with several populations (2018).arXiv Optimization and Control. Available: https://arxiv.org/abs/1810.00783
16. B. Djehiche, A. Tcheukam, H. Tembine, Mean-field-type games in engineering. AIMS Electron. Electri. Eng. **1**(1), 18–73 (2017)
17. J. Barreiro-Gomez, H. Tembine, A tutorial on mean-field-type games and risk-aware controllers. Ann. Rev. Control **50**(1), 317–334 (2020)

Chapter 3
A Survey of Mean Field Game Applications in Wireless Networks

This chapter focuses on the applications of mean field game (MFG) and mean-field-type game (MFTG) in the next generations of wireless networks from 5G (year 2020-) to 6G (year 2030-). These wireless technologies include ultra-dense networks (UDNs), device-to-device (D2D) communications, Internet-of-Things (IoT), unmanned-aerial-vehicle- (UAV-) assisted wireless networks, and mobile edge networks (MENs).

UDNs are networks with high cell density or number of cells per coverage area. The main advantages of UDN over traditional networks are high cell density, idle cell mode capabilities, advanced interference management and frequency reuse, backhaul management, and prevalent line-of-sight (LOS) transmissions. MFG/MFTG have been applied in interference management, propagation modeling, energy efficiency, and scheduling.

D2D communication is a technology that allows nearby mobile devices to communicate directly rather than indirectly through cellular base stations or core network. Meanwhile, IoT is the interconnection of large number of heterogeneous smart devices through the Internet. MFG/MFTG have been utilized in interference management, power control, proximity services, and network security.

UAVs are flying vehicles known for their mobility, maneuverability, and flexibility in applications. UAVs have assisted in expanding the capabilities of wireless networks by serving as aerial base stations, aerial network users, flying ad-hoc network (FANET) access points, and backhaul for terrestrial networks. MFG/MFTG have been realized in channel modeling, energy efficiency, and deployment.

MENs bring network functions closer to the end users through edge networks, which effectively reduces the latency due to the proximity of the edge network to the end users. Applications and use cases of MENs include dynamic content delivery, augmented reality, computation assistance, video streaming and analysis, IoT, connected vehicles, and wireless big data analytics. MFG/MFTG have been envisioned in latency reduction, energy efficiency, resource management, caching, and computation offloading.

© The Editor(s) (if applicable) and The Author(s), under exclusive license to
Springer Nature Switzerland AG 2021
R. A. Banez et al., *Mean Field Game and its Applications in Wireless Networks*,
Wireless Networks, https://doi.org/10.1007/978-3-030-86905-2_3

3.1   Ultra-Dense Networks

3.1.1   Overview of Ultra-Dense Networks

Ultra-dense networks (UDNs) are networks with high cell density or number of cells per coverage area. Networks with cell density greater than 10^3 cells/km^2 are considered ultra-dense [1]. Small cells in UDN can be classified as either base stations (BSs) or access points (APs) [2]. BSs can perform the tasks of a macrocell with lower power over a smaller coverage area. Meanwhile, APs are installed to extend the coverage area of a macrocell. The main advantages of UDN over traditional networks are high cell density, idle cell mode capabilities, advanced interference management and frequency reuse, backhaul management, and prevalent line-of-sight (LOS) transmissions [3]. Please refer to Chap. 1 and the references therein for additional information about UDNs.

Performance metrics are utilized in research works in order to evaluate the performance of UDNs which can lead to strategies that improve the capabilities of UDNs. In [3], metrics used to evaluate the performance of UDNs are discussed. To evaluate the quality of coverage experienced by a user, success probability and rate coverage are typically measured. Success probability refers to the probability that the signal-to-interference-plus-noise ratio (SINR) of any user is above a threshold for good network connection, whereas rate coverage denotes the probability that the achievable rate of any user is above a prescribed minimum rate. For evaluating the frequency spectrum usage of the UDN, the average spectral efficiency is measured, which refers to the average number of transmitted bits per unit bandwidth. In UDNs, high spectral efficiency is achievable because of frequency reuse techniques. Meanwhile, the advantage of high cell density in UDNs is measured by network throughput, which refers to the average data transmission rate per unit bandwidth per unit area [4]. The more dense the UDN, the higher is the network throughput. To assess the power consumption of the UDN, energy efficiency, or the ratio of the network throughput to the power consumption per unit area, is usually observed. Lastly, a fairness index is used to indicate the fairness of resource allocation scheme among the users such that they are treated or serviced equally by the UDN [5].

3.1.2   Research Opportunities and Challenges

The authors of [3] compiled a list of features of UDNs that need further research. The list includes user association, small cell discovery, interference management, backhauling, energy efficiency, and propagation modeling.

The user-to-cell association in UDNs is more challenging than the user association in traditional networks because of dense network structure, stronger inter-cell interference, user mobility, and association of a user to multiple cells in UDNs. A related issue is small cell discovery of a user. A user's ability to detect a small cell becomes more challenging as the interference among the neighboring cells becomes higher. Consequently, interference management is a significant feature of UDNs that justifies the use large number of small cells or BS. It not only deals with reducing the interference among nearby cells but also the interference within a small cell. However, as the number of cells become large, interference management becomes more challenging. Consequently, a good mathematical model is crucial in modeling and analyzing the interference model of UDNs. Managing the backhaul of UDNs is also challenging due to the huge amount of data coming in and out of the BSs. A proposed solution to this problem is using wireless links such as mmWaves and massive-MIMO as backhaul links that can handle huge amount of data transmissions. Another important issue in UDN that must be addressed is energy management. Specifically, the power consumption of small cells must be optimized so that enough power is available for reliable transmission and coverage as well as for mitigated interference against neighboring cells. Finally, UDNs require a robust propagation model that takes into account not only a prevalent LOS signal but also the fading introduced by multi-path signals.

3.1.3 Proposed Mean Field Game Solutions

3.1.3.1 Interference Management

Interference management in UDNs requires a model of the strength of the signal interference experienced by a generic or representative base station (or end user equipment). UDNs typically consist of a large number of small cell base stations where each base station may suffer interference from a group of interfering base stations. An exact model that characterizes the interference experienced by any base station would require complete information of its interferers. With MFGs, an approximate model of interference is sufficient to be able to analyze UDNs and implement interference management techniques.

Consider a UDN with a set $\mathcal{N}$ of base stations. The interference experienced by a transmission from a base station $i \in \mathcal{N}$ to a user k is given by

$$I_{ik}(p_{i'}, t) = \sum_{i' \in \mathcal{N} \setminus \{i\}} p_{i'}(x, t) h_{i'k}(t), \tag{3.1}$$

where transmit power $p_{i'}$ represents the transmit power of other base stations $i' \in \mathcal{N} \setminus \{i\}$ and $h_{i'k}$ denotes the channel gain from base station i' to user k.

In [6], assuming the base stations are identical, the interference experienced by a transmission from a base station to a user is reformulated as a mean-field interference $\mu(m(t), t)$,

$$\mu(m(t), t) = \eta \int_{\mathcal{X}} p(x, t) h(t) m(x, t) \, dx, \tag{3.2}$$

where η is a normalization factor, $p(x, t)$ denotes the transmit power of a base station based on its state x at time t, $h(t)$ refers to the channel gain, and $m(x, t)$ refers to the mean field of the state.

For a two-tier network consisting of a group of macro base stations, $\mathcal{M}$, as leaders and a group of small base stations, $\mathcal{N}$, as followers, the authors of [7] remodeled the mean-field interference using stochastic geometry. Any small base station $i \in \mathcal{N}$ experiences interference $I_{ii'}$ from other base stations and interference I_{ij} from the macro base stations $j \in \mathcal{M}$, Mathematically,

$$
\begin{aligned}
I_{ii'}(p_{i'}, t) &= \sum_{i' \in \mathcal{N} \backslash \{i\}} p_{i'}(t) h_{ii'}(t), \\
I_{ij}(p_j, t) &= \sum_{j \in \mathcal{M}} p_j(t) h_{ij}(t),
\end{aligned}
\tag{3.3}
$$

where the interference $I_{ii'}(p_{i'}, t)$ and $I_{ij}(p_j, t)$ are functions of the control of base stations other than base station i. Using stochastic geometry, the interference $I_{ii'}(p_{i'}, t)$ and $I_{ij}(p_j, t)$ have been substituted respectively by the mean-field interference $\mu_s(m(t), \lambda_s)$ and $\mu_m(p_m, \lambda_m)$,

$$
\begin{aligned}
\mu_s(m(t), \lambda_s) &= \mathbb{E}[I_{ii'}(p_{i'}, t)] = 2\pi \lambda_s \int_{\mathscr{X}} p(x, t) m(x, t) \, dx, \\
\mu_m(p_m, \lambda_m) &= \mathbb{E}[I_{ij}(p_j, t)] = 2\pi p_m \lambda_m,
\end{aligned}
\tag{3.4}
$$

where p_m and λ_m denote the transmit power and density of macro base stations, and λ_s refers to the density of small base stations. The mean-field interference $\mu_s(m(t), \lambda_s)$ refers to the mean-field interference experienced by a base station from other base stations, while $\mu_m(p_m, \lambda_m)$ refers to the mean-field interference experienced from the macro base stations. Note that these interference are not functions of specific parameters of other base stations.

3.1.3.2   Propagation Modeling

The signal strength at a receiver is affected by the state of the channel in which the signal was transmitted. Wireless channels are usually affected by path loss due to the distance between a transmitter and a receiver and fading due to the multi-path copies of the same signal that arrive at a receiver. In wireless networks, the channel dynamics $dh(t)$ is usually described by the stochastic differential equation

$$
dh(t) = f(x, m, t) \, dt + \sigma(t) \, dw(t),
\tag{3.5}
$$

where the deterministic term $f(x, m, t) \, dt$ refers to the average channel gain due to path loss and $\sigma \, dw(t)$ refers to the fading channel gain due to multi-path.

The quantity $w(t)$ represents a standard Wiener process. Assuming the channel dynamics follows the Ornstein-Uhlenbeck dynamics [8],

$$dh(t) = 0.5(\mu_h - h(t))\,dt + \sigma_h^2\,dw(t), \tag{3.6}$$

where $h(t)$ refers to the mean-field approximation of the aggregate interference channel state, μ_h and σ_h^2 denote its mean and variance.

In [9], power control and traffic offloading in UDNs with a dominating player is investigated. A dominating player refers to a base station with much higher interference contribution compared to other base stations in the network. The dynamics of the interference state μ_0 of the dominating base station is given by

$$d\mu_0(t) = p_i(t)dh_0(t) = 0.5p_i(t)(\mu_{h_0} - h_0(t))\,dt + \sigma_{h_0}^2(t)\,dw(t), \tag{3.7}$$

while the dynamics of the interference state μ_i of a generic base station is

$$d\mu_i(t) = p_0(t)dh_{0i}(t) + 0.5p_{i'}(t)(\mu_{h_i} - h_i(t))\,dt + \sigma_{h_i}^2(t)\,dw(t), \tag{3.8}$$

where the mean-field approximation $h_0(t)$ denotes the aggregate effect of the smaller base stations to the channel dynamics of the dominating base station and the mean-field approximation $h_i(t)$ refers to the aggregate effect of base stations $i' \in \mathcal{N} \setminus \{i\}$ to base station i.

Another method of modeling the propagation complexities in UDNs is through a robust mean-field optimization. In robust MFG, an optimal control is found by optimizing the cost under the worst-case disturbance in state dynamics. In [10], an interference-aware power control using robust MFG was studied. Assuming the dynamics of the interference state μ_i of a generic base station has an additional disturbance $\xi_i(t)$ due to uncertainties in the interference state μ_i, the robust stochastic game problem encountered by base station i becomes

$$p_i^*(t) = \arg\min_{p_i} \max_{\xi_i} \mathbb{E}[J(x_i, p_i, \xi_i)]. \tag{3.9}$$

where the cost functional is defined as

$$J_i(x_i, p_i, \xi_i) = \int_0^T r_i(x_i, p_i, t)\,dt + r_i(T) - \rho^2 \int_0^T \xi_i^2(t)\,dt. \tag{3.10}$$

Note that in the absence of disturbance, the cost functional reduces to $J_i(x_i, p_i) = \int_0^T r_i(x_i, p_i, t)\,dt + r_i(T)$ and optimization problem reduces to $p_i^*(t) = \arg\min_{p_i} \mathbb{E}[J(x_i, p_i)]$.

3.1.3.3 Energy Efficiency

In the most general sense, energy efficiency refers to how much work is done per unit of energy spent. In wireless networks, energy efficiency has been defined as the ratio of the total network throughput in bits per second and the total power consumption of the network in watts, $\eta_e = T/P$, [11].

Consider a UDN with N base stations. If a base station i serves M end users, its energy efficiency is given by

$$\eta_{ei}(t) = \frac{\sum_{j=1}^{M} C_{ij}(p_i, p_{i'}, t)}{p_{Ti}(t)} = \frac{\sum_{j=1}^{M} B_{ij} \log_2 \left(1 + \frac{p_i(t)h_{ij}(t)}{I_{ij}(p_{i'}, t) + \xi^2}\right)}{p_{Ti}(t)}, \qquad (3.11)$$

where C_{ij} denotes the throughput from base station i to end user j and $p_{Ti}(t)$ refers to the total power consumption of base station i. The interference term $I_{ij}(p_{i'}, t)$ corresponds to the interference experienced by end user j from base stations other than base station i, whereas ξ^2 refers to the additive noise received by the end user.

Numerous research have been accomplished to improve the energy efficiency of UDNs while satisfying a target QoS for the end users. For instance, the authors of [12] proposed an MFG framework to optimize the energy efficiency in 5G full-duplex cellular networks such that interference is mitigated while managing the SINR requirement for quality transmission. The proposed work allows base station i to find the transmit power that maximizes the weighted sum of the mean uplink energy efficiency and the mean downlink energy efficiency. These efficiencies are derived as

$$\eta_{ei}^{DL}(t) = \frac{\Phi_1(\bar{p}_i, m(x, t), \lambda_b)}{p_{Ti}(t)},$$
$$\eta_{ei}^{UL}(t) = \frac{\Phi_2(\bar{p}_i, m(x, t), \lambda_b)}{p_{Tj}(t)}, \qquad (3.12)$$

where $\Phi_1(\bar{p}_i, m(x, t), \lambda_b)$ and $\Phi_2(\bar{p}_i, m(x, t), \lambda_b)$ are the derived mean throughput functions for downlink and uplink transmissions, respectively, which are functions of the average transmit power $\bar{p}_i$, the mean field of network energy efficiency $m(x, t)$, the density of base stations λ_b, and the total power consumption $p_{Ti}(t)$ of base station i and $p_{Tj}(t)$ of end user j. The mean energy efficiencies do not depend on the knowledge of the transmit powers of other base stations.

3.1.3.4 Scheduling

The massive number of end users in a dense network not only poses interference and network traffic issues but also scheduling challenges. With many users trying to access the same network resources at the same time, a scheduling algorithm is necessary to avoid severe network congestion.

In [13], the authors proposed an algorithm based on stochastic geometry and MFG that computes optimal transmission powers of the base stations such that the consumed energy is minimized while also completing the user requests they received. In [6], the authors developed a scheduling algorithm using MFG and Lyapunov optimization. The algorithm requires the derived mean field interference in order to calculate the optimal transmission powers for the base stations.

3.1.4   Summary

Table 3.1 summarizes works on UDN discussed in the previous subsection as well as other similar works in the literature that proposed an MFG-based approach. Aside from the traditional MFG framework that consists of a pair of HJB/FPK equations, other MFG forms such as robust MFG (RMFG) and MFG with a dominating player (MFGDP) and mathematical tool such stochastic geometry (SG) have also been used.

Table 3.1 Summary of MFG-based research works on UDN

Research direction	Ref.	Model	Optimize	Mean-field Use	State	Control
Interference management	[10]	RMFG	SINR level	Mean SINR level	Interference dynamics	BS transmit power
	[7]	MFG	Interference and SINR level	Mean transmit power	Energy level	BS transmit power
Energy efficiency	[6]	MFG	Energy efficiency	Mean-field interference	Queue state and channel gain	BS transmit power and EU schedule
	[12]	MFG	Downlink/uplink energy efficiency	Mean-field interference	Energy level	BS transmit power
Propagation modeling	[9]	MFGDP	Interference and SINR level	Mean interference and SINR level	Interference dynamics	BS transmit power
Scheduling	[13]	MFG, SG	Energy efficiency	Mean-field interference	Queue state and channel gain	EU transmit power
Mobility management	[14]	MFG, SG	Energy efficiency	Mean-field interference	BS location distribution	EU transmit power

3.2 Device-to-Device Communications and Internet-of-Things

3.2.1 Overview of Device-to-Device Communications and Internet-of-Things

Device-to-device (D2D) communication is a technology that allows nearby mobile devices to communicate directly, rather than indirectly, through cellular base stations or core network [15]. There are four types of D2D communications according to [16]: operator-controlled D2D relay links, operator-controlled direct D2D communications, device-controlled D2D relay links, and device-controlled direct D2D communication. The link establishment of operator-controlled links is controlled by the operator, while the link establishment of device-controlled links is controlled by the source and destination devices. Some of the key applications of D2D include traffic offloading, emergency services, cellular coverage extension, data dissemination, reliable health monitoring, and mobile tracking and positioning [17].

Internet of Things (IoT) is the interconnection of large number of heterogeneous smart devices through the Internet. It allows smart devices to communicate, make decisions, and deliver services together [18]. An IoT network has three main components: sensors/devices as data sources, IoT gateways for data pre-processing, and cloud/core network for data processing [19]. The emerging IoT applications include smart home, intelligent transportation system, smart city, industrial, and smart healthcare [20]. In order to successfully implement these applications, the key requirements include low deployment cost, long battery life, low device cost, extended coverage area, support for massive number of connected devices (scalability), and security and privacy. Please refer to Chap. 1 and the references therein for additional information on D2D and IoT.

3.2.2 Research Opportunities and Challenges

In [17] and [21], issues on the implementation of D2D communication include interference management, power control, mode selection, network or device discovery, proximity and context-aware services, mobility, network coding, and network security. Communications between D2D devices suffer from D2D intercell and intracell interferences as well as interference from cellular network users. These disturbances can be managed by interference avoidance, interference coordination, and interference cancellation. Moreover, these disturbances can be optimized by power control and mode of operation selection of the devices. Network or device discovery refers to the ability of D2D devices to identify other nearby devices. Because of limited power supply, network discovery should be done as energy efficient as possible. Meanwhile, proximity services such as health monitoring, device-to-vehicle communications, and social networks need more investigation to provide users with better experience and network operators with new business opportunities. Finally, networking coding is essential in D2D communication to improve the success of end-to-end data delivery, while network security is important to ensure secure sharing of sensitive information between D2D devices. The involvement of D2D communications in vehicular communications, mmWave spectrum band, social networks, and energy harvesting also require further research.

Meanwhile, the major challenges that occur in IoT includes availability, reliability, mobility, performance, management, scalability, interoperability, security, and privacy [18]. The IoT must be available to provide services to any IoT devices using IoT applications. In addition, the IoT devices must be compatible with the functionalities and protocols of the IoT. The IoT should be able to maintain reliable success of IoT service delivery including mobile users. Performance evaluation of IoT services and proper management of the interconnection of the devices are necessary to improve the services provided by the IoT. The IoT should also be able to add new devices, services, and functions without sacrificing the integrity of the network. The IoT faces interoperabilitiy issues because of the interconnection of large number of heterogeneous devices. Lastly, IoT should be able to guarantee secure sharing of information and the privacy of its users.

3.2.3 Proposed Mean Field Game Solutions

3.2.3.1 Interference Management and Power Control

In order to enhance the performance of D2D networks, the interference present on D2D links between the devices must be mitigated.

Given the mean-field interference μ_i experienced by D2D link i, the power control problem is finding the optimal power p_i^* that minimizes the cost associated with the mean-field interference while still passing the minimum SINR threshold,

$$p_i^*(t) = \arg\min_{p_i} \mathbb{E}[J(x_i, p_i, \mu_i)]. \tag{3.13}$$

where $J(x_i, p_i, \mu_i)$ is the cost functional of D2D link i that depends on the mean-field interference μ_i and the SINR level at D2D link i.

In [22], interference in D2D networks is handled by controlling the transmit power of the devices. The interference and power control is formulated as an MFG problem where each device finds its optimal power that minimizes the interference and power consumption costs. While interference management is important, the transmit power of the devices must be sufficient in order to establish quality links with other devices. In [23], energy efficiency of the IoT devices is improved by reducing the interference that each device causes. The problem is formulated as an MFG where the optimal power is calculated by optimizing the utility function that includes the cost due to power consumption and payoff from establishing quality communication links.

3.2.3.2 Proximity Services

In [24], the authors proposed a cooperation scheme between D2D devices that stream the same popular contents. The proposed algorithm allows these D2D devices to stream part of the contents and share them with other D2D devices. The problem is formulated as an MFG where the optimal allocation of content between the devices is computed by minimizing the deficit of the value of the content allocation among the devices (i.e., the assignment of parts of the contents to the devices must be optimal).

3.2.3.3 Network Security

Battling the spread of misinformation in IoT networks is investigated in [25]. The problem is formulated as an MFG where each device aims to minimize the cost associated with misinformation [i.e., the quality-of-information (QoI)] and the damage and influence of misinformation (i.e., security risk and reach). The goal is to seek the acceptance probabilities of the information while minimizing the QoI (i.e., quantify the possible risk contained in a piece of information).

3.2.4 Summary

Table 3.2 summarizes the works on D2D and IoT discussed in the previous subsection that utilized an MFG framework to model a research problem or issue.

Table 3.2 Summary of MFG-based research works on D2D and IoT

Research Direction	Ref.	Model	Optimize	Mean-field Use	State	Control
Interference management	[22]	MFG	SINR level and power consumption	Mean-field interference	Interference dynamics and energy level	Device transmit power
	[23]	MFG	SINR level and power consumption	Mean-field interference	Energy level	Device transmit power
Proximity services	[24]	MFG	Value of content allocation	Distribution of states	Content length and deficit queue length	Content allocation
Network security	[25]	MFG	Quality-of-Information	Distribution of states	Condition of node based on received information	Information acceptance probability

3.3 Unmanned Aerial Vehicle Networks

3.3.1 Overview of Unmanned Aerial Vehicle Networks

Unmanned aerial vehicles (UAVs), also known as drones, are flying vehicles known for its mobility, maneuverability, and flexibility in applications. UAVs can be classified according to payload, flying mechanism, range and altitude, speed and flight time, and power supply [26]. UAVs have played major roles in recent advances in communications, civil, and military applications. UAVs in communication networks can be utilized either as aerial base stations to extend the coverage, increase the capacity, and improve the reliability and energy efficiency of wireless communication networks or as mobile access points to enable applications such as real-time video streaming and content delivery [27]. Please refer to Chap. 1 and the references therein for additional information regarding UAVs.

Applications of UAVs in wireless communications have gotten tremendous attention because of the benefits and enhancements that UAVs offer to wireless networks. In [27], the authors discussed the key contributions of UAVs in wireless networks which include enhancing the performance of terrestrial networks, assisting IoT communications involving massive number of devices and sensors, and forming a multi-UAV flying ad-hoc network (FANET). Other UAV applications in wireless networks include backhaul for terrestrial networks and wireless data center for smart cities.

3.3.2 Research Opportunities and Challenges

According to [27] and [28], research directions in UAVs that need further investigation include channel modeling, 3D deployment, trajectory optimization, performance analysis, network planning, resource management, energy efficiency, and security and privacy.

Integrating UAVs in wireless networks complicates channel modeling because the air-to-ground channel (A2G) characteristics have to be taken into account as well, in contrast with traditional terrestrial networks that are only affected by the ground channel characteristics. Consequently, aside from UAV-to-ground communications, channel modeling in wireless networks with UAVs has to consider UAV-to-UAV communications if the network has multiple UAVs and UAV-to-satellite communications if the network is supported by satellites. The 3D deployment or positioning of UAV in UAV-based communications networks has to be optimized because it affects the functionalities the UAVs are assigned. For example, optimized positioning of the UAVs is critical in maximizing the coverage area and network capacity. Path or trajectory planning is important for UAVs because optimized path or trajectory improves energy efficiency, time management, and safety of the UAVs. Performance analysis is crucial in evaluating the significance of UAVs in wireless networks and improving the services provided by the networks. Network planning is crucial in the implementation of a wireless network with optimal number of UAVs, efficient frequency planning, dynamic UAV deployment, mobile drone-EUs, and signaling overhead analysis. Resource management and energy efficiency are required to dynamically manage network resources such as bandwidth, transmit power, and the number of deployed UAVs while controlling the energy consumption of the UAVs. Finally, security and privacy must be given proper attention because UAVs are vulnerable to attack due to the absence of human controller.

3.3.3 Proposed Mean Field Game Solutions

3.3.3.1 Channel Modeling

In UAV-assisted wireless networks, channel modeling is challenging because of line-of-sight (LOS) and non-line-of-sight (NLOS) paths between the UAVs and the terrestrial end users. The proposed work in [29] utilized MFG in order to manage the interference between the UAVs and the terrestrial end users. The MFG problem is calculating the velocity of the UAVs that optimizes the cost due to SINR and flight energy consumption. The cost is expressed in terms of a mean-field interference, which a function of the distribution of the altitudes of the UAVs and the channel condition.

3.3.3.2 Energy Efficiency

The authors of [30] proposed an energy-efficient downlink transmission for multi-UAV networks. The problem is formulated as an MFG that aims to calculate the transmit power of a UAV that maximizes its energy efficiency given the existing interference between the UAVs. The interference has been expressed as a mean-field interference so that the optimization does not require complete information of the states of the UAVs. The resulting MFG problem has been solved using a deep reinforcement learning algorithm.

3.3.3.3 Deployment

Optimized deployment or maneuvering of UAVs is crucial in UAV-based wireless networks in order to maintain the wireless network coverage of the end users while the UAVs fly on limited energy resources. In [31], the authors proposed an MFG-based framework and deep reinforcement learning solution to UAV control for efficient wireless communication coverage. The MFG problem requires the flight speed that maximizes a coverage score given the mean-field distribution of the locations of the UAVs. Then, the MFG problem has been solved using a deep reinforcement learning algorithm.

3.3.3.4 Trajectory Optimization

Trajectory optimization has been important in UAV-assisted wireless networks in order to allow safe and collision-free maneuvering of UAVs while they are performing their designated tasks. In [32], collision-free trajectory for UAVs in sensing applications has been proposed using MFG. The UAVs must be able to do their designated sensing tasks while optimizing their energy consumption. In addition, the UAVs are evaluated using successful task probability for task assignments and channel capacity for quality communication. The work is extended to multiple populations in [33] where the UAVs are owned by different operators.

3.3.4   Summary

Table 3.3 summarizes works on UAV-assisted wireless networks discussed in the previous subsection as well as other similar works in the literature that utilized an MFG framework. Aside from the traditional MFG framework that consists of a pair of HJB/FPK equations, other mathematical frameworks that have also been utilized are multiple population MFG (MPMFG) and machine learning techniques such as deep reinforcement learning (DRL), neural networks (NN), and federated learning (FL).

3.4   Mobile Edge Networks

3.4.1   Overview of Mobile Edge Networks

In mobile edge networks (MENs), network functions that are usually executed at core or cloud networks are brought closer to the end users through edge networks. MENs effectively reduce latency because of the proximity of the edge network to the end users while still being able to deliver high throughput to the end users. MENs have four main architectures: mobile edge computing (MEC), fog computing, cloudlet, and edge caching [39]. MEC, currently known as multi-access edge computing [40], is a standard that allows the computation and storage of computation- and data-intensive tasks at an edge network located near the end users instead of executing them at a cloud or core network which may be located far from the end users. It is an evolution from mobile cloud computing (MCC) which is designed to compute and store of computation- and data- intensive tasks from the mobile devices. With MEC, the edge networks located in proximity with the end users are given the computation and storage capabilities of cloud or core networks. The advantages of MEC include reduced congestion in the core and backhaul network, low latency associated with shorter distance between the edge network

Table 3.3 Summary of MFG-based research works on UAV-assisted wireless networks

Research direction	Ref.	Model	Optimize	Mean-field Use	State	Control
Channel modeling	[29]	MFG	SINR level and energy consumption	Mean SINR and energy consumption	UAV altitude	UAV flight speed
	[34]	MFG, DRL	Network throughput	Mean-field interference	Elevation and azimuth positions	Beam position and power consumption
Energy efficiency	[30]	MFG, DRL	Energy efficiency	Distribution of UAV transmit powers	Interference and energy level	UAV transmit power
Deployment	[31]	MFG, DRL	Energy consumption	Distribution of UAV controls	UAV position, energy consumption, and coverage area	UAV flight direction and flight distance
	[35]	MFG	Energy consumption and UAV position	Distribution of UAV locations	UAV location	Transformation of UAV location
	[36]	MFG	Energy consumption and collision-free flocking	Distribution of UAV locations	UAV location	UAV velocity
Trajectory optimization	[32]	MFG	Energy consumption and collision avoidance	Distribution of UAV locations	UAV position or location	UAV velocity
	[33]	MPMFG	Energy consumption and collision avoidance	Distribution of UAV locations	UAV position or location	UAV velocity
	[37]	MFG, NN	Flight time, energy consumption, and collision avoidance	Distribution of UAV states	UAV location and velocity	UAV acceleration
Communication	[38]	MFG, FL	Flight time, energy consumption, and collision avoidance	Distribution of UAV states	UAV location and velocity	UAV acceleration

and the end users, high bandwidth through the utilization of the edge network, and access to real-time information. Please refer to Chap. 1 and the references therein for additional information regarding MENs.

Because of these benefits from MEC, end users can have access to a variety of MEC applications such as computation offloading, dynamic content delivery and caching, web performance enhancements, augmented reality and video streaming, and support of big data analysis and Internet-of-Things (IoT) [39, 41]. Computation offloading is an MEC application in which an end user equipment, such as a mobile device, can offload computation- and data-intensive tasks to the edge network for execution due to the limited computing capability of the mobile device. MEC also allows dynamic content delivery and caching. For instance, popular contents may be cached at the edge network so that increased demand for the content does not result in interruptions. In addition, MEC supports web performance enhancement by providing contents based on user data, fast web browsing, and quick loading due to edge caching. It allows real-time applications such as video streaming and augmented reality because of low latency and high bandwidth features of edge networks. Moreover, MEC can assist in the computation, storage, and real-time analysis of big data as well as perform data aggregation and big data analytics to IoT data.

3.4.2 Research Opportunities and Challenges

While MEC is a promising solution to the challenges encountered in MCC, advanced research is required to fully take advantage of its benefits. Specifically, issues on computation offloading, latency, energy efficiency, resource management, storage, and security need further investigations [42, 43].

Mobile devices benefit from computation offloading by transferring the computation- and data-intensive tasks to the nearby edge network. However, determining which devices have to offload at a given time and how much task to offload are problems that need to be addressed in order to produce offloading decisions that benefit the users as well as the edge network. Consequently, offloading decisions must take into account constraints such as the latency associated with transmission and execution of data and the energy consumption of the mobile devices. Moreover, proper management and control of the MEC resources must be implemented such that the computing and storage resources are efficiently utilized while satisfying the QoS and QoE requirements of the end users. Finally, security of the data and users must be assured since many end users have access to the edge network.

3.4.3 Proposed Mean Field Game Solutions

3.4.3.1 Latency and Energy Consumption Optimization

Assume that a computation task to be executed at a mobile device has a length of L_m CPU cycles. If f_m is the mobile device CPU computing capability in CPU cycles/second, the computation time and energy consumption associated with executing the computation task at the mobile device CPU are defined as

$$\tau_m = \frac{L_m}{f_m},$$
$$E_m = \kappa_m L_m f_m^2, \tag{3.14}$$

where κ_m is a constant that depends on the architecture of the mobile device CPU [44]. Similarly, if a computation task of length L_s CPU cycles is to be executed at an MEC server with computing capability of f_s in CPU cycles/second, the total delay and energy consumption associated with executing the computation task at the MEC server are defined as

$$\tau_s = \tau_t + \tau_q + \frac{L_s}{f_s},$$
$$E_s = \kappa_s L_s f_s^2, \tag{3.15}$$

where τ_t is the round trip transmission time, τ_q is the queuing delay, and κ_s is a constant that depends on the architecture of the MEC server CPU [44].

Although edge networks can improve the delay that can otherwise be incurred from utilizing cloud or core networks, significant demand from end users may increase the delay associated with edge network utilization. Hence, UAVs have been proposed as solution to reduce latency or delay. The authors in [45] proposed a robust MFG formulation for multi-UAV assisted caching in edge networks. The MFG problem is computing the distance a UAV can fly such that it minimizes the cost associated with caching delay, energy consumption, and disturbance in content popularity and flight condition of the UAV.

3.4.3.2 Resource Management

While edge network has evolved from cloud networks to bring cloud-network capabilities closer to end users, cloud-based networks continue to be an attractive technology for large-scale computing and storage applications. Managing the shared resources of edge and cloud networks is paramount in achieving the QoS of the end users and clients. Specifically, developing efficient and fair resource sharing algorithms is essential in edge and cloud network resource management. In [46], MFG has been applied to manage resource sharing in cloud-based networks. The

MFG problem is finding the optimal strategy or partition of cloud resources, such as CPU access or (last level cache) LLC memory, so that the payoff of each client is maximized. The payoff is a function of the mean-field of the strategy of the cloud network clients.

3.4.3.3 Caching

To fully take advantage of edge caching, popular contents are stored at the edge network in order to reduced latency and traffic congestion at the core network. In the literature, the popularity of a content has been modeled according to Zipf's law and the Chinese Restaurant Process (CRP). According to Zipf's law, the relative probability of a request for the ith most popular content is proportional to $1/i^\alpha$ [47],

$$P_N(i) = \frac{\left(\sum_{i=1}^{N} \frac{1}{i^\alpha}\right)^{-1}}{i^\alpha} \tag{3.16}$$

where $0 < \alpha \leq 1$ is the Zipf parameter. Meanwhile, the mean popularity μ of a content can be modeled through CRP [48]. In CRP, the users take turns requesting for their content. Popular contents have higher probability of getting requested than new contents [49]. If the content request probability dynamics is modeled by the Ornstein-Uhlenbeck process, then

$$dx(t) = a(\mu - x)\,dt + \sigma\,dw(t), \tag{3.17}$$

where a is a chage rate parameter and μ is the mean popularity computed from the CRP model.

A popular method of managing the storage of cloud-based networks is through edge caching in which the most popular contents are cached or stored at the edge network to reduce the delay related to accessing a cloud network. In [48], an edge caching problem has been formulated as an MFG problem in which the base stations calculate the amount of content to be cached that minimizes the cost due to backhaul and overhead costs. Meanwhile, this cost is expressed in terms of the mean-field interference between the base stations and the overlap between popular contents. Meanwhile, the authors of [50] studied distributed edge caching in fog radio access networks. In this work, MFG is utilized to find the caching policy of each fog access point that minimizes the cost due to the request service delay and the traffic load at fronthaul links.

Table 3.4 Summary of MFG-based research works on edge networks

Research direction	Ref.	Model	Optimize	Mean-field use	State	Control
Delay or Latency Optimization	[45]	RMFG	Cache delay and flight energy consumption	Distribution of UAV controls	Flight distance and content request probability	UAV flight speed
Resource management	[46]	MFG	Cache capacity	Distribution of resource allocation decisions	Cloud resource usage	VM resource allocation decision
	[52]	MFG, DRL	Task delay and energy consumption	Distribution of states	SINR and channel gain	BS resource allocation decision
Storage	[48]	MFG	Backhaul and storage capacity	Interference and content overlap	Content request probability and storage capacity	BS transmit power
	[50]	MFG	Request service delay and fronthaul traffic load	Distribution of cache state	File cache state	AP cache decision
Computation offloading	[51]	MFG, DRL	Service delay and energy consumption	Distribution of users	Number of users in a server	Server transition distribution of users
	[53]	MFG, DRL	Total expected response time	Distribution of jobs	Number of jobs in a server	Server job proportion

3.4.3.4 Computation Offloading

Computation offloading allows mobile devices with limited resources, such as computing power and energy, the ability to offload computation tasks to the MEC server. The decision of a mobile device to execute the computation task locally or to offload the computation task to the MEC server depends on which mode of execution will incur the smaller cost. The cost usually depends on execution time and energy consumption.

MFGs have also found applications in computation offloading in edge networks. The authors of [51] proposed an MFG-based task allocation algorithm that allows mobile devices to offload tasks to multiple edge servers. The MFG problem of a mobile device is finding the optimal task allocation to edge devices so that the cost incurred by a mobile device due to execution latency and energy consumption is minimized.

3.4.4 Summary

Table 3.4 summarizes works on MENs discussed in the previous subsection as well as other similar works in the literature that implemented MFG-based algorithms to solve a research problem. Aside from the traditional MFG framework that consists of a pair of HJB/FPK equations, robust MFG (RMFG) and deep reinforcement learning (DRL) have also been utilized.

References

1. M. Ding, D. López-Pérez, G. Mao, P. Wang, Z. Lin, Will the area spectral efficiency monotonically grow as small cells go dense? in *Proceedings of the IEEE Global Communications Conference (GLOBECOM)*, San Diego, CA (2015), pp. 1–7
2. D. López-Pérez, et al., Enhanced intercell interference coordination challenges in heterogeneous networks. IEEE Trans. Wirel. Commun. **18**(3), 22–30 (2011)
3. M. Kamel, W. Hamouda, A. Youssef, Ultra-dense networks: a survey. IEEE Commun. Surv. Tutorials **18**(4), 2522–2545 (2016)
4. C. Li, J. Zhang, K.B. Letaief, Throughput and energy efficiency analysis of small cell networks with multi-antenna base stations. IEEE Trans. Wirel. Commun. **13**(5), 2505–2517 (2014)
5. R. Jain, D.-M. Chiu, W.R. Hawe, *A Quantitative Measure of Fairness and Discrimination for Resource Allocation in Shared Computer System*, Eastern Res. Lab., Digit. Equipment Corp., Hudson (1984)
6. S. Samarakoon, M. Bennis, W. Saad, M. Debbah, M. Latva-aho, Ultra dense small cell networks: turning density into energy efficiency. IEEE J. Select. Areas Commun. **34**(5), 1267–1280 (2016)
7. P. Semasinghe, E. Hossain, Downlink power control in self-organizing dense small cells underlaying macrocells: a mean field game. IEEE Trans. Mobile Comput. **15**(2), 350–363 (2016)

8. T. Alpcan, H. Boche, M.L. Honig, H.V. Poor, *Mechanisms and Games for Dynamic Spectrum Allocation: Reacting to the Interference Field* (Cambridge University Press, Cambridge, 2014)

9. Y. Zhang, C. Yang, J. Li, Z. Han, Distributed interference-aware traffic offloading and power control in ultra-dense networks: mean field game with dominating player. IEEE Trans. Vehicular Technol. **68**(9), 8814–8826 (2019)

10. C. Yang, H. Dai, J. Li, Y. Zhang, Z. Han, Distributed interference-aware power control in ultra-dense small cell networks: a robust mean field game. IEEE Access **6**, 12608–12619 (2018)

11. T. Chen, H. Kim, Y. Yang, Energy efficiency metrics for green wireless communications, in *Proceedings of the International Conference in Wireless Communications and Signal Processing (WCSP)*, Suzhou (2010), pp. 1–6

12. X. Ge, H. Jia, Y. Xiao, Y. Li, B. Vucetic, Energy efficient optimization of wireless-powered 5G full duplex networks: a mean field game approach. IEEE Trans. Green Commun. Netw. **3**(2), 455–467 (2019)

13. M. de Mari, E.C. Strinati, M. Debbah, T.Q.S. Quek, Joint stochastic geometry and mean field game optimization for energy-efficient proactive scheduling in ultra dense networks. IEEE Trans. Cognit. Commun. Netw. **3**(4), 766–781 (2017)

14. J. Park, S.Y. Jung, S.-L. Kim, M. Bennis, M. Debbah, User-centric mobility management in ultra-dense cellular networks under spatio-temporal dynamics, in *Proceedings of the IEEE Global Communications Conference (GLOBECOM)*, Washington (2016), pp. 1–6

15. Y.-D. Lin, Y.-C. Hsu, Multihop cellular: a new architecture for wireless communications, in *Proceedings of the IEEE INFOCOM*, vol. 3 (2000), pp. 1273–1282

16. M.N. Tehrani, M. Uysal, H. Yanikomeroglu, Device-to-device communication in 5G cellular networks: challenges, solutions, and future directions. IEEE Commun. Mag. **52**, 86–92 (2014)

17. F. Jameel, Z. Hamid, F. Jabeen, S. Zeadally, M.A. Javed, A survey of device-to-device communications: research issues and challenges. IEEE Commun. Surv. Tutor. **20**(3), 2133–2168 (2018)

18. A. Al-Fuqaha, M. Guizani, M. Mohammadi, M. Aledhari, M. Ayyash, Internet of things: a survey on enabling technologies, protocols, and applications. IEEE Commun. Surv. Tutor. **17**(4), 2347–2376 (2015)

19. W. Yu, F. Liang, X. He, W.G. Hatcher, C. Lu, J. Lin, X. Yang, A survey on the edge computing for the internet of things. IEEE Access **6**, 6900–6919 (2017)

20. G.A. Akpakwu, B.J. Silva, G.P. Hancke, A.M. Abu-Mahfouz, A survey on 5G networks for the internet of things: communication technologies and challenges. IEEE Access **6**, 3619–3647 (2017)

21. R. Ansari, C. Chrysostomou, S.A. Hassan, M. Guizani, S. Mumtaz, J. Rodriguez, J.J.P.C. Rodrigues, 5G D2D networks: techniques, challenges, and future prospects. IEEE Syst. J. **12**(4), 3970–3984 (2018)

22. C. Yang, J. Li, P. Semasinghe, E. Hossain, S.M. Perlaza, Z. Han, Distributed interference and energy-aware power control for ultra-dense D2D networks: a mean field game. IEEE Trans. Wirel. Commun. **16**(2), 1205–1217 (2017)

23. L. Li, Y. Xu, Z. Zhang, J. Yin, W. Chen, Z. Han, A prediction-based charging policy and interference mitigation approach in the wireless internet of things. IEEE J. Sel. Areas Commun. **37**(2), 439–451 (2019)

24. J. Li, R. Bhattacharyya, S. Paul, S. Shakkottai, V. Subramanian, Incentivizing sharing in realtime D2D streaming networks: a mean field game perspective. IEEE/ACM Trans. Netw. **25**(1), 3–17 (2017)

25. N. Abuzainab, W. Saad, A multiclass mean-field game for thwarting misinformation spread in the internet of battlefield things. IEEE Trans. Commun. **66**(12), 6643–6658 (2018)

26. A. Fotouhi, H. Qiang, M. Ding, M. Hassan, L.G. Giordano, A. Garcia-Rodriguez, J. Yuan, Survey on UAV cellular communication: practical aspects, standardization advancements, regulation, and security challenges. IEEE Commun. Surv. Tutor. **21**(4), 3417–3442 (2019)

27. M. Mozaffari, W. Saad, M. Bennis, Y.H. Nam, M. Debbah, A tutorial on UAVs for wireless networks: applications, challenges, and open problems. IEEE Commun. Surv. Tutor. **21**(3), 2334–2360 (2019)

28. B. Li, Z. Fei, Y. Zhang, UAV communications for 5G and beyond: recent advances and future trends. IEEE Int. Things J. **6**(2), 2241–2263 (2019)
29. L. Li, Z. Zhang, K. Xue, M. Wang, M. Pan, Z. Han, AI-aided downlink interference control in dense interference-aware drone small cells networks. IEEE Access **8**, 15110–15122 (2020)
30. L. Li, Q. Cheng, K. Xue, C. Yang, Z. Han, Downlink transmit power control in ultra-dense UAV network based on mean field game and deep reinforcement learning. IEEE Trans. Vehic. Technol. **69**(12), 15594–15605 (2020)
31. D. Chen, Q. Qi, Z. Zhuang, J. Wang, J. Liao, Z. Han, Mean field deep reinforcement learning for fair and efficient UAV control. IEEE Int. Things J. **8**(2), 813–828 (2021)
32. Y. Kang, S. Liu, H. Zhang, W. Li, Z. Han, S. Osher, H.V. Poor, Joint sensing task assignment and collision-free trajectory optimization for mobile vehicle networks using mean-field games. IEEE Int. Things J. **8**(10), 8488–8503 (2021)
33. Y. Kang, S. Liu, H. Zhang, Z. Han, S. Osher, H.V. Poor, Task selection and collision-free route planning for mobile crowd sensing using multi-population mean-field games. IEEE Trans. Green Commun. Netw. (Early Access, 2021)
34. L. Li, H. Ren, Q. Cheng, K. Xue, W. Chen, M. Debbah, Z. Han, Millimitere-wave networking in the sky: a machine learning and mean field game approach for joint beamforming and beam-steering. IEEE Trans. Wirel. Commun. **19**(10), 6393–6408 (2020)
35. K. Xue, Z. Zhang, L. Li, H. Zhang, X. Li, A. Gao, Adaptive coverage solution in multi-UAVs emergency communication system: A discrete mean field game, in *Proceedings of the 14th International Wireless Communications & Mobile Computing Conference (IWCMC)*, Limassol (2018), pp. 1-6
36. H. Kim, J. Park, M. Bennis, S.-L. Kim, Massive UAV-to-ground communication and its stable movement control: A mean-field approach, in *Proceedings of the IEEE 19th International Workshop on Signal Processing Advances in Wireless Communications (SPAWC)*, Kalamata (2018), pp. 1–6
37. H. Shiri, J. Park, M. Bennis, Massive autonomous UAV path planning: A neural network based mean-field game theoretic approach, in *Proceedings of the IEEE Global Communications Conference (GLOBECOM)*, Waikoloa (2019), pp. 1–6
38. H. Shiri, J. Park, M. Bennis, Communication-efficient massive UAV online path control: federated learning meets mean-field game theory. IEEE Trans. Commun. **68**(11), 6840–6857 (2020)
39. S. Wang, X. Zhang, Y. Zhang, L. Wang, J. Yang, W. Wang, A survey on mobile edge networks: convergence of computing, caching, and communications. IEEE Access **5**, 6757–6779 (2017)
40. I. Morris, *ETSI Drops "Mobile" From MEC*, Light Reading, New York (2016)
41. T. Taleb, K. Samdanis, B. Mada, H. Flinck, S. Dutta, D. Sabella, On multi-access edge computing: a survey of the emerging 5G network edge cloud architecture and orchestration. IEEE Commun. Surv. Tutor. **19**(3), 1657–1681 (2017)
42. N. Abbas, Y. Zhang, A. Taherkordi, T. Skeie, Mobile edge computing: a survey. IEEE Int. Things J. **5**(1), 450–465 (2018)
43. P. Mach, Z. Bacvar, Mobile edge computing: a survey on architecture and computation offloading. IEEE Commun. Surv. Tutor. **19**(3), 1628–1656 (2017)
44. W. Zhang, Y. Wen, K. Guan, D. Kilper, H. Luo, D. Wu, Energy-optimal mobile cloud computing under stochastic wireless channel. IEEE Trans. Wirel. Commun. **12**(9), 4569–4581 (2013)
45. L. Li, K. Xue, Q. Cheng, D. Wang, W. Chen, M. Pan, Z. Han, Delay optimization in multi-UAV edge caching networks: a robust mean field game. IEEE Trans. Vehic. Technol. **70**(1), 808–819 (2021)
46. A.F. Hanif, H. Tembine, M. Assaad, D. Zeghlache, mean-field games for resource sharing in cloud-based networks. IEEE/ACM Trans. Netw. **24**(1), 624–637 (2016)
47. L. Breslau, P. Cao, L. Fan, G. Phillips, S. Shenker, Web caching and zipf-like distributions: evidence and implications, in *IEEE INFOCOM*, New York (1999), pp. 1–9
48. H. Kim, J. Park, M. Bennis, S.-L. Kim, M. Debbah, Mean-field game theoretic edge caching in ultra-dense networks. IEEE Trans. Vehic. Technol. **69**(1), 935–947 (2020)

49. E. Bastug, M. Bennis, M. Debbah, Living on the edge: the role of proactive caching in 5G wireless networks. IEEE Commun. Mag. **52**(8), 82–89 (2014)
50. Y. Jiang, Y. Hu, M. Bennis, F.-C. Zheng, X. You, A mean field game-based distributed edge caching in fog radio access networks. IEEE Trans. Commun. **68**(3), 1567–1580 (2020)
51. D. Shi, H. Gao, L. Wang, M. Pan, Z. Han, H.V. Poor, Mean field game guided deep reinforcement learning for task placement in cooperative multiaccess edge computing. IEEE Int. Things J. **7**(10), 9330–9340 (2020)
52. L. Li, Q. Cheng, X. Tang, T. Bai, W. Chen, Z. Ding, Z. Han, Resource allocation for NOMA-MEC systems in ultra-dense networks: a learning-aided mean field game approach. IEEE Trans. Wirel. Commun. **20**(3), 1487–1500 (2021)
53. H. Gao, W. Li, R.A. Banez, Z. Han, H.V. Poor, Mean field evolutionary dynamics in dense-user multi-access edge computing systems. IEEE Trans. Wirel. Commun. **19**(12), 7825–7835 (2020)

Chapter 4
Mean Field Game Applications in Ultra-Dense 5G, 6G, and Beyond Wireless Networks

Ultra-dense networks (UDNs) are networks with high number of interacting communication devices and/or equipment per coverage area. The advantages of implementing UDNs are high network capacity, advanced interference management, frequency reuse, and prevalent line-of-sight (LOS) transmissions. Meanwhile, the challenges in UDNs include interference management, energy efficiency, and scheduling. Recently, mean-field game (MFG) theory has been introduced in the economics and engineering literature to study the strategic decision making of large number of symmetric, indistinguishable, and negligible interacting agents. In this chapter, three case studies are presented to demonstrate how MFG was utilized in addressing the challenges in ultra-dense wireless networks and improving their performance.

The first case study deals with a distributed power control method for ultra-dense D2D communications underlying cellular communications. The power control method not only considers the remaining battery energy of the D2D transmitter but also the effects of both the interference caused by the generic D2D transmitter to others and the interference from all others introduced to the generic D2D receiver. The interference among the D2D links is formulated as a mean-field game (MFG) theoretic framework by interference mean-field approximation. The cost function combines both the performance of the D2D communication and cost for transmit power at the D2D transmitter. The Hamilton–Jacobi–Bellman (HJB) and Fokker-Planck–Kolmogorov (FPK) equations of the MFG system are derived.

The second case study focuses on a downlink power control problem of maximizing the energy efficiency in an ultra-dense unmanned aerial vehicles (UAV) network. The power control problem is formulated as a discrete MFG to imitate the interactions among a large number of UAVs, and then the MFG framework is transformed into a Markov decision process (MDP) to obtain the equilibrium solution of the MFG due to the dense deployment of UAVs. Specifically, a deep reinforcement learning-based MFG (DRL-MFG) algorithm is proposed to suppress

R. A. Banez et al., *Mean Field Game and its Applications in Wireless Networks*,
Wireless Networks, https://doi.org/10.1007/978-3-030-86905-2_4

the interference and maximize the energy efficiency by using deep neural networks (DNN) to explore the optimal power strategy for UAVs.

The third case study provides a framework for load balancing in user-dense multi-access edge networks. The study considers two cases: a single-resource case in which the server can only provide one type of resource to the user, such as computation; and a multi-resource case in which the server can provide multiple types of resources to the user, such as computation and storage. To achieve high quality of service (QoS) and low latency under these two cases, mean field evolutionary mechanisms are proposed to solve the load balancing problems under the two cases.

4.1   Introduction

With the advancement of telecommunication technologies, wireless networking (e.g., 5G, 6G, and Beyond) has become ubiquitous owing to the great demand of pervasive mobile applications. The convergence of computing, communications, and media will allow users to communicate with each other and access any content at anytime, anywhere. Future wireless networks will support various services such as high-speed access, telecommuting, interactive media, video conferencing, real-time Internet games, e-business ecosystems, smart homes, automated highways, and disaster relief. Yet many technical challenges remain to be addressed in order to make this wireless vision a reality. A critical issue is devising distributed and dynamic algorithms for ensuring a robust network operation over time-varying and heterogeneous environments. Therefore, in order to support tomorrow's wireless services, it is essential to develop efficient mechanisms that provide an optimal cost-resource-performance tradeoff and that constitute the basis for next generation ubiquitous and autonomic wireless networks.

Game theory is a formal framework with a set of mathematical tools to study the complex interactions among interdependent rational players. Mean field was first studied in physics for the behavior of systems with large numbers of negligible individual particles. Recently, mean-field game (MFG) theory was introduced in the economics and engineering literature to study the strategic decision making by small interacting agents of huge populations. In this chapter, three case studies that utilized MFG are presented.

First, device-to-device (D2D) communications provide significant performance enhancement in terms of spectrum and energy efficiency by proximity and frequency reuse. However, such performance enhancement is largely limited by mutual interference and energy availability, in particular, in ultra-dense D2D networks. In Sect. 4.2, we consider both interference dynamics and available energy of the generic device, and then we formulate an MFG theoretic framework with the interference mean-field approximation. Different from previous works, in addition to the remaining energy state of the battery, we investigate the effects of both the interference dynamics of the generic device introduced to others, and all others' interference dynamics introduced into the generic device on power control. We formulate the cost function by combining both the performance and cost for transmit power. Within the MFG framework, we derive the related Hamilton-Jacobi-Bellman (HJB) and Fokker-Planck-Kolmogorov (FPK) equations. Then, a novel energy and interference aware power control policy is proposed, which is a joint finite difference algorithm based on the Lax-Friedrichs scheme and the Lagrange relaxation to solve the coupled HJB and FPK equations of the corresponding MFG, respectively. The numerical results are presented to verify the spectrum and energy efficiency improvement of our proposed approach.

Second, as an emerging technology in 5G, ultra-dense unmanned aerial vehicles (UAVs) network can significantly improve the system capacity and networks coverage. However, it is still a challenge to reduce interference and improve energy efficiency (EE) of UAVs. In Sect. 4.3, we investigate a downlink power control problem to maximize the EE in an ultra-dense UAV network. Firstly, the power control problem is formulated as a discrete MFG to imitate the interactions among a large number of UAVs, and then the MFG framework is transformed into a Markov decision process (MDP) to obtain the equilibrium solution of the MFG due to the dense deployment of UAVs. Specifically, a deep reinforcement learning-based MFG (DRL-MFG) algorithm is proposed to suppress the interference and maximize the EE by using deep neural networks (DNN) to explore the optimal power strategy for UAVs. The numerical results show that the UAVs can effectively interact with the environment to obtain the optimal power control strategy. Compared with the benchmarks algorithms, the DRL-MFG algorithm converges faster to the solution of MFG and improves the EE of UAVs. Moreover, the impact of the transmit power on EE under the different heights of the UAVs is also analyzed.

Third, multi-access edge computing (MEC) can use the distributed computing resources to serve the large numbers of mobile users in the next generation of communication systems. In this new architecture, a limited number of mobile edge servers will serve a relatively large number of mobile users. Heterogeneous servers can provide either single resource or multiple different resources to the massive number of selfish mobile users. To achieve high quality of service (QoS) and low latency under these two cases, in Sect. 4.4, we construct two system models and formulate our problems as two non-cooperative population games. Then we apply our proposed mean field evolutionary approach with two different strategy graphs to solve the load balancing problems under those two cases. Finally, to evaluate the performance of our algorithms, we employ the following performance indicators: overall response time (average response time of the whole system), individual response time (response time of each server), and fairness index (equality of users' response time).

4.2 Ultra-Dense Device-to-Device Networks

Device-to-device (D2D) communications underlying conventional cellular networks improve energy and spectrum efficiency. These beneficial opportunities are
achieved by proximity and frequency reuse [1]. However, these benefits also come
with technical challenges. Interference management is one of the most critical
challenges for D2D communication. Both intra-tier and inter-tier interferences exist
in D2D communications, which affect the system performance, and thus need to be
mitigated to further improve spectrum efficiency.

Enhancing energy efficiency is another important problem due to the promising
awareness of the energy saving and the environmental protection. Traditional D2D
devices are powered by batteries, therefore, extending the battery life and saving
energy is important to improve users' experience. In summary, the performance
enhancement is largely limited by the interference and high energy consumption, in
particular, in the ultra-dense deployment scenarios [2, 3]. As a result, to optimize
both spectrum and energy efficiency, different techniques have been designed to
mitigate interference and save energy [3–5]. For instance, interference coordination
[3], interference mitigation [4], and resource management [5] have been investigated
aiming at improving spectral and energy efficiency.

Power control is critical to both save energy and mitigate interference for
D2D communications [6–10]. To characterize the dynamically interactive behaviors
due to the coupled interference relationship, game theory has been well applied
to model the resource competition and interference coordination, analyze the
strategic behaviors, and design distributed algorithms. Recently, both cooperative
game and non-cooperative game theory have found extensive applications in D2D
communications [11–19]. However, these classical game models are difficult to
analyze and solve when the number of D2D links becomes very large. Mean filed
games (MFGs) are promising alternatives to model and analyze such a large scale
system.

While the classical games model the interaction of a single player with each of
the other players of the system, an MFG models individual player's interaction with
the average effect of the collective behavior of the players [20–22]. This collective
behavior is modeled by the mean field. Interactions among individual players with
the mean field are modeled by the Hamilton-Jacobi-Bellman (HJB) equation in the
mean field game. And, the dynamics of the mean field according to the players'
actions can be modeled by a Fokker-Planck-Kolmogorov (FPK) equation. These
coupled FPK and HJB equations are also called forward and backward equations,
respectively. The mean field equilibrium of an MFG can be obtained by solving
these two equations. MFGs have found their applications in many fields, such as
cognitive radio networks [23], green power control [24], hyper-dense heterogeneous
networks [25, 27], cloud-based networks [28], and smart power grids [29]. Different
from the existing literature, for a dense D2D network we pursue a distributed power
control policy for a long term period, $[0, T]$, where T can be hundreds of frames, and
we consider both the available energy at the D2D transmitters and the interference
dynamics in the network into consideration.

In this case study, we present a rational power control policy design that can help save energy and mitigate interference. On the one hand, more rational power use means battery life can last longer. On the other hand, all players implement power control with the impact of interference on other players taken into consideration. This is a win-win situation in which the wireless environment has lower interference. The remainder of this section is organized as follows. In Sect. 4.2.1, we present the system model and formulate the problem for ultra-dense D2D power control. The corresponding D2D differential game is presented in Sect. 4.2.2. Finally, some pointers about the equivalent optimal control problem of the differential game and the distributed iterative algorithm to obtain the MFG equilibrium of the problem are stated in Sect. 4.2.3. For more details about this work, please refer to [30].

4.2.1   System Model and Problem Formulation

We consider the uplink of an ultra-dense D2D communications network. The D2D communication pairs share uplink resources with some existing macrocell user equipments (MUEs). We assume that there are N D2D pairs sharing the same channel with the cellular uplink, as shown in Fig. 4.1. The term "ultra-dense" implies the following: (1) the number of D2D communication pairs is very large, and that is $N \to +\infty$; (2) full frequency reuse, i.e., the D2D communication pairs use the same frequency; and (3) most of the user devices select the D2D communication mode, and there exists a relatively small number of macrocell users.

In the network, both intra-tier and inter-tier interference exist. For instance, as shown in Fig. 4.1, the D2D transmitter D2D_1^T communicating with its receiver introduces intra-tier interference to other D2D receivers. At the same time, due to full frequency reuse, D2D_1^T causes inter-tier interference link to the MUE. Here, we

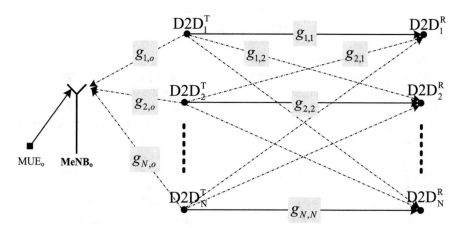

Fig. 4.1 A D2D network with a large number of D2D links

define the interference introduced by player i, $i \in \mathcal{N}$ to others $j \in \mathcal{N}$, $j \neq i$ at time t as

$$I_{i \to}(t) = \sum_{j=1, j \neq i}^{N} p_i(t) g_{i,j}(t), \qquad (4.1)$$

where $p_i(t)$ is the transmit power corresponding to D2D pair i, $i \in N$, and $g_{i,j}(t)$ defines the channel gain from the D2D pair i's transmitter to the D2D pair j's receiver, $j \in N$, $j \neq i$. Therefore, (4.1) gives the interference introduced by player i, $i \in N$ to all other D2D receivers at time t, where player i, $i \in N$ is called the generic D2D transmitter.

At the same time, the transmission of player i, $i \in N$ also introduces interference to player o, where we define the only existing uplink MUE to Macrocell evolved node B (MeNB) pair as player o. We define the inter-tier interference introduced by player i to player o as

$$I_{i \to o}(t) = p_i(t) g_{i,o}(t), \qquad (4.2)$$

where $g_{i,o}(t)$ is channel gain between D2D pair i to macrocell link o.

Finally, the interference perceived by the D2D pair i at time t, which is the interference introduced by other D2D links to the generic D2D link i, is given as

$$I_{\to i}(t) = \sum_{j=1, j \neq i}^{N} p_j(t) g_{j,i}(t). \qquad (4.3)$$

Here, we assume that orthogonal channels are used for different MUEs, and we do not consider any power control policy at the macrocell layer. Our focus is on the power control policy for the D2D transmitters.

The achieved signal-to-interference-plus-noise ratio (SINR) at the receiver of D2D pair i at time t is

$$\gamma_i(t) = \frac{p_i(t) g_{i,i}(t)}{I_{\to i}(t) + \sigma^2}, \qquad (4.4)$$

where σ^2 is the thermal noise power.

With the above definition of SINR, the power control problem can be summarized as follows: each player i will determine the optimal power control policy $Q_i^\star(0 \longrightarrow T)$ with the interference $I_{i \to}$ introduced to others, the interference introduced by others $I_{\to i}$, and remaining energy. The D2D transmitters adapt their transmit power during the time interval of $t \in [0, T]$. This power control problem can be formulated as a differential game due to the interference dynamics and the energy dynamics [20–25]. In this differential game, the interference from other D2D links to the generic D2D receiver will be considered in the cost function while the interference introduced by the generic D2D transmitter to other D2D links will be regarded as one of the two state variables. The other state variable will be the remaining energy at the generic D2D transmitter.

4.2.2 Differential Game Model for Power Control

The differential game model for power control in the D2D network described above is defined as follows:

Definition 4.1 The D2D differential power control game G_s for D2D transmitters is defined by a 5-tuple: $G_s = (\mathcal{N}, \{\mathcal{P}_i\}_{i \in \mathcal{N}}, \{\mathcal{S}_i\}_{i \in \mathcal{N}}, \{\mathcal{Q}_i\}_{i \in \mathcal{N}}, \{c_i\}_{i \in \mathcal{N}})$, where

- **Player set** $\mathcal{N}$: $\mathcal{N} = \{1, ..., N\}$ represents the player set of densely-deployed D2D communication pairs. They are rational policy makers in the D2D game, where the number N of the D2D links is huge, and even goes to infinity.
- **Set of actions** $\{\mathcal{P}_i\}_{i \in \mathcal{N}}$: This is the set of possible transmit powers. Each transmitter determines the power $p_i(t) \in \{\mathcal{P}_i\}$ at any time $t \in [0, T]$ to minimize the cost function (to be defined later).
- **State space** $\{\mathcal{S}_i\}_{i \in \mathcal{N}}$: We define the state of player i as the combination of the interference introduced by the D2D transmitter i to other D2D links and the remaining energy at this D2D transmitter. The state space is composed of all possible states.
- **Control policy** $\{\mathcal{Q}_i\}_{i \in \mathcal{N}}$: a full power control policy is denoted by $\mathcal{Q}_i(t)$, with $t \in [0, T]$ to minimize the average cost over the time interval T with two dimensional states.
- **Cost function** $\{c_i\}_{i \in \mathcal{N}}$: we will define a novel cost function, where we consider both the achieved performance, e.g., the SINR and the transmit power.

To determine the control policy $\{\mathcal{Q}_i\}_{i \in \mathcal{N}}$, we need to define the state space $\{S_i\}_{i \in \mathcal{N}}$ and the cost function $\{c_i\}_{i \in \mathcal{N}}$.

4.2.2.1 State Space

The power control policy $\mathcal{Q}_i(t)$, for $t \in [0, T]$ of player i, $i \in \mathcal{N}$ is determined to find the optimal control policy which minimizes the cost subject to given state dynamics. The state space is defined based on the intra-tier and inter-tier interferences in (4.1) and (4.2), respectively, and the energy usage dynamics.

Energy Usage Dynamics The remaining energy state $E_i(t)$ at time t equals to the amount of available energy. Meanwhile, $0 \leq E_i(t) \leq E_i(0)$, where $E_i(0)$ is the energy at time 0. The power control actions at time t should be any $p_i(t) \in [0, p_{max}]$, where p_{max} is the maximum possible transmit power. Without loss of generality, we define the evolution law of the remaining energy in the battery as

$$dE_i(t) = -p_i(t)dt, \tag{4.5}$$

which means that energy $E_i(t)$ of the battery decreases with the transmit power consumption $p_i(t)$. At the same time, in game G_s, each player i should also consider the impact of interference on other players.

Interference Dynamics With intra-tier and inter-tier interference defined in (4.1) and (4.2), respectively, we first define the interference function that describes the interference caused by the generic D2D transmitter to others as

$$\mu_i(t) = I_{i\rightarrow}(t) + I_{i\rightarrow o}(t), \tag{4.6}$$

where (4.6) describes all the interference impacts introduced by player i to other D2D pairs $j \in \mathcal{N}$, $j \neq i$ and the only MUE o. According to definitions in (4.1) and (4.2), we have

$$\mu_i(t) = \sum_{j=1, j\neq i}^{N} p_i(t)g_{i,j}(t) + p_i(t)g_{i,o}(t). \tag{4.7}$$

To simplify the notation, we further represent (4.7) as

$$\mu_i(t) = p_i(t)\varepsilon_i(t), \tag{4.8}$$

where $\varepsilon_i(t) = \sum_{j=1, j\neq i}^{N} g_{i,j}(t) + g_{i,o}(t)$. From (4.8), the total interference at time t to others depends on $p_i(t)$ and $\varepsilon_i(t)$ at time t. Therefore, we can define the interference state as

$$d\mu_i(t) = \varepsilon_i(t)dp_i(t) + p_i(t)\partial_t\varepsilon_i(t). \tag{4.9}$$

Meanwhile, the mean-field approximation method to estimate the channel gains $\varepsilon_i, i \in \mathcal{N}$ is discussed in [30].

We define the following state space for player i:

$$s_i(t) = [E_i(t), \mu_i(t)], \quad i \in \mathcal{N}, \tag{4.10}$$

where the interference caused by the generic D2D transmitter to other D2D links is regarded as one of the state variables. Also, the other state variable is the remaining energy as given in (4.5). The power control policy $\mathcal{Q}_i(t)$, for $t \in [0, T]$ should minimize the cost function $c_i(t)$, which will be defined in the next subsubsection, with consideration of the remaining energy state $E_i(t)$ and the interference state $\mu_i(t)$.

The interference state $\mu_i(t)$ in (4.9) of the generic D2D transmitter will affect the strategy of the player, and all others' interference $I_{\rightarrow i}(t)$ introduced to the generic receiver will affect the SINR performance. To distinguish between these two interferences, we further clarify them as State$_1$ and State$_2$, respectively, as shown in Fig. 4.2.

In Fig. 4.2, the interference dynamics introduced by the generic D2D link to other D2D links is regarded as State$_1$, which is one of the state variable. Also, we define the State$_2$ as the interference dynamics from other D2D links to the generic D2D link, which we will combine the effect of State$_2$ in the defined cost function, and State$_1$ as the constrained conditions.

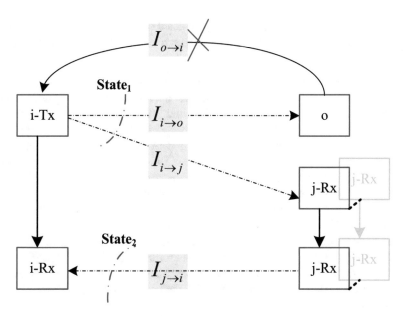

Fig. 4.2 Interference dynamics in an ultra-dense D2D network. $State_1$: Interference introduced by the generic D2D transmitter to other D2D links; $State_2$: Interference from other D2D links to the generic D2D link

4.2.2.2   Cost Function

With the above definition of state space $s_i(t)$, each D2D transmitter i will determine the optimal power control policy $\mathscr{Q}_i^{\star}(t)$, with $t \in [0, T]$ to minimize the cost. Generally, the communication performance is related to the SINR definition $\gamma_i(t)$ in (4.4), and we also introduce the identical SINR threshold γ_{th}. Here, γ_{th} is predefined to meet the communication requirements. Therefore, the cost function is given by

$$c_i(t) = (\gamma_i(\mathrm{t}) - \gamma_{th}(\mathrm{t}))^2 + \lambda p_i(t), \tag{4.11}$$

where λ is introduced to balance the units of the achieved SINR difference and the consumed power. Here, it is clear that the player i will minimize the SINR difference from the threshold and the power consumption at any time t. It is easy to prove that the cost function $c_i(t)$, given by (4.11) is convex with respect to $p_i(t)$.

4.2.3   Optimal Control Problem and Mean Field Equilibrium

In the previous subsections, we formulate an MFG theoretic framework for ultra-dense D2D networks, where we assume that the number of D2D links can approach infinity. In this framework, we jointly consider the remaining energy at the D2D transmitters and the interference as the state space and obtain an optimal distributed power control policy. In [26], we have formulated the optimal control problem and study the MFG equilibrium, which are summarized as follows. For more details, please refer to [26].

1. **Energy and interference-aware problem formulation:** In the proposed MFG framework, the problem is formulated as a cost minimization problem with two kinds of interference into consideration. The interference from other D2D links to the generic D2D link is investigated in the cost function, while the interference dynamics introduced by the generic D2D transmitter to other D2D links is regarded as one of the constraints. Also, the other constraint condition is the remaining energy level at the D2D transmitters. For the effects of both the types of interferences, we propose a mean-field approximation approach. This facilitates designing a distributed power control policy for a generic D2D transmitter. Moreover, this leads to a social optimal power control.
2. **Distributed iterative algorithm to obtain the MFG equilibrium:** We derive the corresponding HJB and FPK equations for the presented D2D MFG framework. A joint finite difference algorithm based on the Lax-Friedrichs scheme and Lagrange relaxation is proposed to solve the coupled HJB and FPK equations, respectively.

In the sequel, some numerical results are presented to illustrate the mean field distributions and the power control policy of a generic D2D transmitter. The proposed algorithm can improve both spectrum and energy efficiency when compared to some benchmark schemes.

The downlink transmission of an OFDMA D2D network, with the radius of D2D links uniformly distributed between 10 to 30 m is considered. We set the system parameters as the bandwidth $w = 20$ MHz, and background noise power is 2×10^{-9} W as the noise power spectral density is $\kappa = -174$ dBm/Hz. Without special instructions, we choose the standardized case with 500 LTE frames, the maximum energy is 0.5 J, the number of D2D links will vary from $N = 50$ to $N = 200$. The path-loss exponent for D2D links is 3. The duration of one LTE radio frame is 10 ms, and for 500 frames, $T = 5$ s. We also pick, E_{max} to be 0.5 J. The tolerable interference level of each player μ_{max} is assumed to be 5.8×10^{-6}.

We describe the three dimensional mean field distributions and power policy, as shown in Fig. 4.3. However, mean field and power policy are four dimensional vectors. Therefore, we plot mean field and power policy for three cases i.e., (a) mean field distributions with varying interference and energy but fixed time; (b) mean field distributions with varying time and energy but fixed interference; and (c) mean field distributions with varying time and interference but fixed energy; similarly, we illustrate the power control policies according to the three cases. Basically, we can see that both the remaining energy and the interference dynamics affect the mean field distributions and power control policy. The result implies that it is important to consider the mean field of the interference and energy consumption of the devices so that an individual device can save energy and mitigate interference by implementing the optimal power control policy based on the mean field.

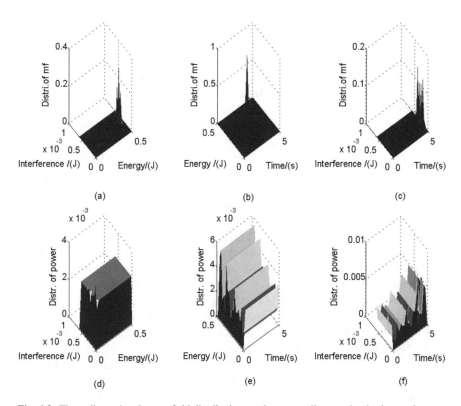

Fig. 4.3 Three dimensional mean field distributions and power policy varying in time and space. For the distributions of mean field we have three cases with (**a**) $T = 5$; (**b**) $\mu_{max} = 5.8 \times 10^{-6}$; (**c**) $E_{max} = 0.1$ fixed, respectively. Similarly, for the distributions of powers we also have three cases with (**d**) $T = 5$; (**e**) $\mu_{max} = 5.8 \times 10^{-6}$; and (**f**) $E_{max} = 0.1$, respectively

4.3 Ultra-Dense Unmanned Aerial Vehicles Networks

With the rapid development of the mobile Internet and the Internet of Things (IoT), ultra-dense networks (UDNs) [31] have become one of the critical technologies in the fifth-generation mobile communication system (5G) [32]. A UDN can meet the requirements of explosive data traffic. It is capable of expanding network communication range and system throughput effectively by densely deploying base stations in limited range [33].

Nowadays, the communication technology of flying platforms such as unmanned aerial vehicles (UAVs) also has become one of the essential technologies in 5G for improving the ability of 5G communication systems with various emergent emergencies. This is because the traditional ground base station (GBS) cannot meet the demands of flexible deployment and rapid recovery in some specific events or emergency scenario nowadays [4]. UAVs have many characteristics such as small sizes, strong portability, and high flexibility, so it can replace the GBSs in special scenarios, such as sea, disaster areas, etc. At present, UAVs have been widely used in the military, civilian, and public fields [34–36].

However, a single UAV often fails to meet the needs of complex tasks in large-scale scenarios due to its limited range of services. In order to solve this problem, it is necessary to design the ultra-dense UAV network. Compared with traditional UAV communication, the ultra-dense UAV network has many advantages, such as significantly improving the networks service range and data transmission rate, increasing the communication capacity, and satisfying access for a large number of users, and so on. However, the ultra-dense UAV network communication still faces enormous challenges. Firstly, the UAV-to-user links need to consider the additional interference and unique channel characteristics different from the traditional ground communication channels. Secondly, it is difficult for the UAVs to work for a long time because the energy of UAVs is limited [37]. Consequently, how to improve the communication performance of UAVs is an urgent problem. To solve this problem, we investigate the downlink power control problem in the ultra-dense UAV network. It optimizes the transmit power of UAVs to improve the energy efficiency (EE) and reduce the interference effectively.

At present, game theory has been widely used in the field of communication, such as solving the power control problem. For UAV networks, each UAV increases its own transmit power as much as possible to improve the communication quality of the users it serves. However, it causes serious interference to the users served by other UAVs. In order to guarantee the communication quality of the served users, other UAVs also increase the transmit power. Therefore, it is necessary to find an equilibrium solution among UAVs to optimize the transmit power to prevent excessive interference, which is in line with the application of game theory. The classical game problems have to consider the interactions among each individual and the other individuals to establish the coupling relationships. However, the coupling relationships will be greatly increased when a large number of individuals is involved, which leads to high computational complexity. Therefore, the power control problems solved with classical game theory methods become very difficult and sometimes impossible to be solved. Based on this, mean field game (MFG) method is proposed [38]. MFG can be considered as a special branch of differential game with a large number of agents. Classical game theory models the interactions of a single rational player with the other players. However, MFG models any individual's selfish behavior with respect to the collective behavior of all the other players. Here, the collective behavior is named as a mass. In the MFG framework, MFG can well explain the interactions between individual behavior and collective behavior, which are described by the Hamilton-Jacobi-Bellman (HJB) equation. Meanwhile, the dynamics evolution of the mean field is described by the Fokker-Planck-Kolmogorov (FPK) equation. The HJB equation is called the backward equation, and the FPK equation is called the forward equation. Then the equilibrium solution of MFG is obtained by solving the two equations.

Usually, the numerical analysis method is used to solve the HJB and FPK equations [39]. However, it has an initialization-sensitive problem for the iteration of the HJB and FPK equations, and the computational complexity is particularly high when the system has large-scale state space and action space. Meanwhile, MFG provides a descriptive modeling method for the UDN, which can reflect how the system finds the optimal control policy [40, 41]. This allows it to be linked to the Markov decision process (MDP) and reinforcement learning (RL) [42]. Moreover, when the number of agents is very large, it is difficult for RL algorithms to handle the interactions among multiple agents in a complex environment in which MFG can be used to simplify the model.

Nowadays, machine learning (ML) has been widely used in the field of wireless communication [43–50]. In [51], the author provided the theoretical guidance for the application of ML in wireless communication networks. RL is an efficient and important learning algorithm in ML. It can build an environment model while exploring the unknown environment, and it learns the mapping from environment to action to get an optimal policy. Deep reinforcement learning (DRL) [52] combines the perception ability of deep learning (DL) with the decision-making ability of RL, and it can directly control according to the input information. However, due to the complexity of the control problem of the ultra-dense UAV network, such as the coverage of the UAV, energy consumption, power allocation and other problems, the traditional RL cannot successfully complete the task. So the DRL method is applied to solve the MFG equilibrium solution in order to solve the problems of high-dimensional state space and huge computational cost.

In this case study, we design an ultra-dense UAV network for the large-scale emergency communication scenarios. In the emergency communication system, a large number of UAVs replace damaged GBSs for air-to-ground service, with dense users and wide areas. However, each UAV is interfered by other UAVs because all the UAVs are sharing the same downlink channel, which causes the UAVs to compete with each other. By adjusting the power, the interference among the UAVs can be effectively reduced to improve the EE of the UAVs. So in order to solve the downlink power control problem in the ultra-dense UAV network, we formulate it as an MFG framework. Then, the cost function of UAV is derived by studying the signal-to-interference-plus-noise ratio (SINR) of the users served by the UAV. In order to solve the high complexity problem of MFG, we reduce the MFG problem into an MDP. Therefore, a DRL based on MFG (DRL-MFG) algorithm is proposed that allows the UAV to predict its value function by exploring the optimal policy.

The rest of the section is organized as follows. In Sect. 4.3.1, the system model of the ultra-dense UAV network is discussed. In Sect. 4.3.2, the associated MFG framework is formulated, and the DRL-MFG algorithm in solving the downlink power control problem of the ultra-dense UAV network is introduced.

4.3.1   System Model

In this subsection, the downlink power control problem in the ultra-dense UAV network is investigated. In an emergency communication system, a large number of UAVs replace damaged GBSs for air-to-ground service with dense users and

Table 4.1 Table of symbols

Symbol	Definition
$\mathcal{M}$	The number of the GBSs
$\mathcal{U}$	The number of the UAVs
$\mathcal{K}$	The number of the users
$p_r(t)$	The receive power of the users
$p_i(t)$	The transmit power of the UAVs
$\lvert X_{i,k}(t) \rvert$	Distance between ith UAV and kth user
$p_m(t)$	The transmit power of GBSs
H	Height of the UAVs
p_c	The fixed circuit power
$p_e(t)$	The consuming power of the UAVs
$E_i(t)$	The energy of the UAVs
$I_{u \to k}(t)$	The interference from other UAVs to user
$I_{m \to k}(t)$	The interference from the GBSs to user
α_u	The path-loss exponent in the UAVs-to-users link
α_m	The path-loss exponent in the GBSs-to-Users link
$\gamma_i^k(t)$	The SINR of the user
$R_i^k(t)$	The throughput of the user
τ_k	The target threshold SINR of the user
N_0	The thermal noise power
η	Energy efficiency
W	The system bandwidth

wide areas. In practice, the spatial distribution of UAVs is random and independent of each other because the UAVs are deployed opportunistically in an unplanned manner as most of the time. Moreover, the UAVs use the position optimization method in literature [53] to obtain the best position, and then hover in the air to provide services to ground users. To improve the frequency band utilization, we assume that all the UAVs adopt the frequency reuse technology, which brings more massive co-channel interference to the users. In order to obtain a real-time optimal power control, it is unrealistic for a centralized allocation strategy to be subject to network overhead under the dense network. Therefore, a distributed scheme based on game theory is considered in which the UAV adjusts its transmit power based on the power control policy. However, these references do not consider the situation when the action is continuous.

We consider the existence of M intact GBSs in the network, which is defined as $\mathcal{M} = \{m_1, m_2, \cdots m_M\}$. In addition, each UAV serves multiple users in the UAVs network. However, for the convenience of description, assuming that each UAV base station serves only one user at a specific time slot, and we set up the ith ($i \in \mathcal{U}$) UAV serves the kth ($k \in \mathcal{K}$) user at time t in the ultra-dense UAV network. It is worth noting that the assumption is established due to the fairness and interchangeability among users. The number of UAVs in the network is defined as $\mathcal{U} = \{u_1, u_2, \cdots, u_U\}$. The number of users in the network is represented as $\mathcal{K} = \{k_1, k_2, \cdots, k_K\}$. The set of transmit power is expressed as

$\mathscr{P} = (\mathscr{P}_M, \mathscr{P}_U)$, where $\mathscr{P}_M = \{p_1, p_2, \cdots, p_M\}$ is the set of transmit power of the GBS, and $\mathscr{P}_U = \{p_1, p_2, \cdots, p_U\}$ is the set of transmit power of the UAVs. Please refer to Table 4.1 for the summary of the symbols and their definitions.

Figure 4.4 illustrates the system model of the ultra-dense UAV network. Firstly, the channel model between the UAV and the user is investigated. When the UAV is hovered above the ground, path-loss and shadow fading are introduced due to the existence of ground obstacles [54]. We assume that the height H of the UAV is a fixed value. It is worth mentioning that different heights of UAVs will affect system performance. The coordinate of the ith UAV in the horizontal domain is denoted by (x_i, y_i). The coordinate of the kth user served by the ith UAV is denoted by $(x_{i,k}, y_{i,k})$. Therefore, the distance between the ith UAV and the kth user can be defined as

$$\left|X_{i,k}(t)\right| = \sqrt{H^2 + (x_{i,k} - x_i)^2 + (y_{i,k} - y_i)^2}. \tag{4.12}$$

As a consequence, the received power of the kth user can be expressed as

$$p_r(t) = p_i(t)g_{i,k}(t)\left|X_{i,k}(t)\right|^{-\alpha_u}, \tag{4.13}$$

where $g_{i,k}(t) = 10^{\frac{v_{i,k}(t)}{10}}$ denotes the shadow effect in path-loss, $v_{i,k}(t)$ is a Gaussian random variable, which is usually between 0–10 dB, and α_u is the path-loss exponent in the UAVs-to-users link.

Considering the energy $E_i(t)$ of the UAV is limited, its available energy can be expressed as $[0, E_i(0)]$, where $E_i(0)$ is the maximum energy at initial time $t = 0$. The evolution of energy is defined by the differential equation

$$dE_i(t) = -p_i(t)dt. \tag{4.14}$$

Considering the two layers of the network using the same spectrum resources, each user will receive interference from the UAVs and the GBSs, respectively.

The UAVs use frequency reuse technology, and so the kth user is interfered by other UAVs. The interference introduced by other UAVs can be written as

$$I_{u\to k}(t) = \sum_{j=1, j\neq i}^{U} p_j(t)g_{j,k}(t)\left|X_{j,k}\right|^{-\alpha_u}. \tag{4.15}$$

Finally, the interference introduced by the GBSs to users at time t is given as

$$I_{m\to k}(t) = \sum_{m=1}^{M} p_m(t)g_{m,k}(t)\left|X_{m,k}\right|^{-\alpha_m}, \tag{4.16}$$

where $p_m(t)$ is the transmit power of the mth GBS, $g_{m,k}(t)$ is the channel gain from the mth GBS to the kth user, $\left|X_{m,k}\right|$ is the distance between the mth GBS and the kth user, and α_m is the path-loss exponent in the GBSs-to-users link.

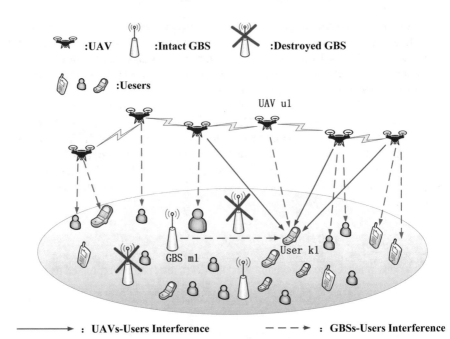

Fig. 4.4 System model of an ultra-dense UAV network for emergency communications

According to (4.15) and (4.16), we define the total interference of the kth user as follows:

$$I_k(t) = I_{u \to k}(t) + I_{m \to k}(t). \tag{4.17}$$

The SINR of user k served by UAV i at time t is given by

$$\gamma_i^k(t) = \frac{P_i(t) g_i(t) \left| X_{i,k}(t) \right|^{-\alpha_u}}{I_{\mu_i(t)} + N_0}. \tag{4.18}$$

The following inequality should hold for any UAV to satisfy its quality of service (QoS) constraint:

$$\frac{P_i(t) g_i(t) \left| X_{i,k}(t) \right|^{-\alpha_u}}{I_k(t) + N_0} \geq \tau_k, \tag{4.19}$$

where τ_k is the target threshold SINR of the kth user.

Therefore, the throughput of the system at time t is

$$R_i^k(t) = W \log_2(1 + \gamma_i^k(t)), \tag{4.20}$$

where W is the system bandwidth.

Meanwhile, we define the transmission efficiency of the energy consumption by the UAV at time t according to (4.18), which can be expressed as

$$\eta = \frac{R_i^k}{p_e(t)} = \frac{W \log_2\left(1 + \gamma_i^k\right)}{p_i(t) + p_c}$$

$$= \frac{W \log_2\left(1 + \dfrac{P_i(t)g_i(t)\left|X_{i,k}(t)\right|^{-\alpha_u}}{I_k(t) + N_0}\right)}{p_i(t) + p_c}, \tag{4.21}$$

where p_c is the fixed circuit power. In [55], the authors had shown that the energy efficiency (EE) is a strict decreasing function of the transmit power when $p_c = 0$, i.e., when the power consumption of the fixed circuit is not considered. However, as $p_c \neq 0$, the EE increases first and then decreases with the increase of transmit power, i.e., the EE is quasi-concave with respect to transmit power.

The downlink power control of UAVs networks can be summarized as follows. Each UAV needs to consider the interference from other UAVs and the other GBSs to users when the UAVs transmit data to users. At the same time, the optimal power control of UAVs can be determined according to the transmission efficiency of energy consumed by UAVs at time t. Here, the power policy is a series of power control actions, which are expressed as the MFG problems.

4.3.2 Mean Field Game Problem Formulation and Analysis

The goal is to maximize the EE of each UAV by controlling its downlink transmit power. The throughput of each user and the received interference of the user will increase with increasing of the UAV transmit power. At the same time, in order to ensure QoS of the users, the transmit power of the UAVs have to be limited. However, each UAV maximizes its EE while also affecting the transmit power of other UAVs. Therefore, this is a non-cooperative game.

Considering the number of UAVs, in this part, we formulate the MFG model to solve the downlink power control problem in the ultra-dense UAV network. Here, we define the interference caused by the ith UAV to other users as

$$I_{i \to \mathcal{K}} = \sum_{f=1, f \neq k}^{\mathcal{K}} p_i(t)g_{i,f}(t)\left|X_{i,f}(t)\right|^{-\alpha_m}. \tag{4.22}$$

According to the system model, the interference state can be defined as

$$dI_{i \to \mathscr{K}} = \sum_{f=1, f \neq k}^{K} g_{i,f}(t) \left| X_{i,f}(t) \right|^{-\alpha_m} dp_i(t). \tag{4.23}$$

Then, we define the state space of the mean field as

$$\pi_i^k(t) = [E_i(t), I_{i \to \mathscr{K}}], \ i \in \mathscr{U}, \tag{4.24}$$

and the distribution of state space can be expressed as

$$\pi^k(t) = [\pi_1^k(t), \ \pi_2^k(t), \ \cdots, \pi_n^k(t)], \tag{4.25}$$

where n is the nth distribution of the state, and N is the size of the distribution of the state space.

The MFG describes the behaviour of the large-scale agents as the mean field term, which is a statistical function characterizing the distribution of the mass. Therefore, the complexity of the system is reduced significantly because a large number of information interactions with other agents are converted into the interaction with the mass. Specifically, in the MFG with a large number UAVs, the fluctuations of UAV's actions are expected to be "average out." Since the impact of other UAV on a single UAV is expressed by the overall actions, the impact of a single UAV on the mean field can be ignored when the deployment of the UAVs is intensive.

The mean field of the ultra-dense UAV network is denoted as $\alpha_{-i}(a^d, t)$, for $d = 1, 2, \cdots, D$, where D is the size of the action space set. Mathematically, it can be written as

$$\alpha_{-i}(a^d, t) = \lim_{U \to \infty} \frac{1}{U} \sum_{i,j \in U, j \neq i} \mathbb{1}(a_{j,t} = a^d), \tag{4.26}$$

where $\mathbb{1}(a_{j,t} = a^d)$ is the indicator function that the action $a_{j,t}$ taken by the jth UAV at time t is a_d. For the sake of convenience, let α represent $\alpha_{-i}(a^d, t)$.

Afterwards, the following assumptions about the UAVs and their interactions with the mean field are made: (1) The number of UAV $\mathscr{U}$ is sufficiently large; (2) Each UAV is homogeneous and exchangeable; (3) Each UAV is interacting with the finite mean field.

It is worth noting that all UAVs are independent, and so the state space of each UAV is the same. The ith UAV determines the reward function based on the remaining energy at the present moment and the interference of the kth user, and then it evolves from one state to another.

In MFG, the HJB equation and FPK equation describe the whole system model. The FPK equation represents the state change of the mean field. It can be expressed as

$$\pi^k(t+1) = \sum_{i=1, i \neq i}^{i,j \in \mathcal{U}} P_{ij}(\alpha, t)\pi^k(t), \tag{4.27}$$

where $P_{ij}(\alpha, t)$ is the transition probability from state i to state j at time t for the kth user under the influence of the mean field α, $i, j \in \mathcal{U}$.

The EE of the ith UAV is defined as the reward function of the model

$$r(t) = \max \eta = \max \frac{W \log_2 \left(1 + \frac{P_i(t)g_i(t)|X_{i,k}(t)|^{-\alpha_u}}{I_k(t) + N_0}\right)}{p_i(t) + p_c},$$

$$s.t. \quad dE_i(t) = -p_i(t)dt, \tag{4.28}$$

$$dI_{i \to \mathcal{K}} = \sum_{j=1, j \neq k}^{K} g_{i,j}(t)|X_{i,j}(t)|^{-\alpha_m} dp_i(t),$$

$$0 \leq p_i(t) \leq p_{\max}, \quad \gamma_i^k \geq \tau_k.$$

However, it can be expressed as the value function,

$$V_i^k = \max_{p_i^t} \{r(t, \pi_i^k(t), p_i(\alpha, t)) + \sum_{i=1, i \neq j}^{i,j \in \mathcal{U}} P_{ij}(t)V_i(t+1)\}. \tag{4.29}$$

From the state equation and the value function, the HJB equation is derived for the optimal control problem that the value function satisfies

$$\frac{\partial V_i^k(t)}{\partial t} - \max_{p_i(t)} \left\{r(t, \pi_i^k(t), p_i(t)) + p_i(t)\frac{\partial V_i^k(t)}{\partial \pi_i^k(t)}\right\} = 0, \tag{4.30}$$

in which the HJB equation simulates the interaction between the individual player and mean field. The equilibrium solution of the MFG can be obtained by solving the FPK and HJB equations.

The HJB function means that the final value of the function is known, and we determine the value of $V_i(t)$ at time $[0, T]$. Therefore, the HJB equation is always solved backwards in time, starting from $t = T$, and ending at $t = 0$. When solved over the entire state space, the HJB equation is a necessary and sufficient condition for the optimality of the UAV control. The FPK equation evolves forward with time. The interactive evolution finally leads to the mean field equilibrium (MFE).

The application of deep reinforcement learning (DRL) to the MFG framework promotes the interactions among the UAVs. The optimal power control strategy of a single UAV is based on the dynamic evolution of the UAV network, whereas the dynamic evolution of the UAV network is updated according to the power control strategy of each UAV. Figure 4.5 shows the sketch of DRL where it can be seen that

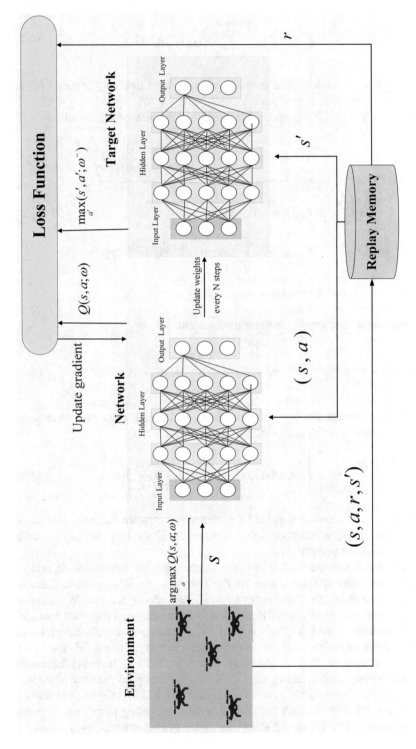

Fig. 4.5 Deep reinforcement learning exhibiting the interaction between the agent and the environment

a deep neural network (DNN) is utilized to learn the optimal power policy for the UAVs. The state and action are the inputs to the DNN in which the hidden layer is a model with a multi-layer neural network structure. The reader may refer to [56] for more details regarding the DRL-MFG algorithm.

4.4   User-Dense Multi-Access Edge Computing Systems

As combination of mobile computing and cloud computing, mobile cloud comput- ing (MCC) has provided variety of services, such as computation and storage, to satisfy the demand of massive mobile users [57]. However, MCC faces significant challenges in the next generation communication networks because the total number of mobile devices continues to grow dramatically with more computation-hungry applications, such as augmented reality (AR). According to the Cisco Visual Networking Index (CVNI) report [58], there will be 12.3 billion mobile-connected devices and the average mobile network connection speed (8.7 Mbps in 2017) will reach 28.5 Megabits per second (Mbps) by 2022. It will be challenging for MCC to satisfy high Quality of Service (QoS) and low latency required by those devices with real-time applications due to frequent uploading and downloading between users and remote core clouds [59]. As a result, new communication architectures are urgently needed.

Multi-access edge computing (MEC) is a promising candidate [60] that can handle the challenges currently being faced by MCC because of the following three characteristics. First, mobile edge hosts are deployed close to mobile devices, which means the physical distance for data transmission will be much shorter than MCC. Second, mobile edge servers can analyze big data in a local manner and run isolated from the rest of the MEC network, which means heavy uploading and downloading traffic generated by millions of mobile users will decrease sharply. Finally, MEC can provide high QoS with ultra low latency and high bandwidth. However, to turn these advantages of MEC into reality, many challenges still remain [61]. For example, in a user-dense MEC network, a limited number of servers are serving a relatively large number of users. These selfish users are only interested in their own profits such as shorter response time, while limited servers are working independently without a central controller. This will result in poor performance of the system. Therefore, an efficient load balancing algorithm is needed to achieve the good performance required by MEC [62].

In this case study, we apply a mean field evolutionary approach to solve the load balancing problem in user-dense MEC for the following reasons: (1) Separately dealing with the interactions between a large number of users will increase the computational complexity significantly for other game theoretic methods, such as the Nash non-cooperative game and traditional evolutionary game [63, 64]. (2) Instead of reacting to other users separately when seeking the optimal strategy for a generic user, a mean field game (MFG) regards them as a mean field and reacts only to the collective behavior [65]. This will reduce the computational complexity

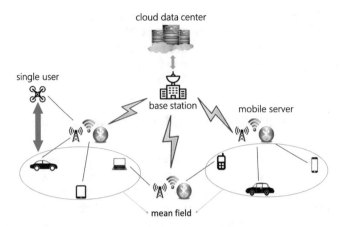

Fig. 4.6 System model of a user-dense MEC for load balancing problem

significantly [66]. (3) We consider both the single-resource case and the multi-resource case of the servers. The single-resource case is the situation when the server can only provide one type of resource to the user, such as computation. The multi-resource case refers to the situation when the server can provide multiple types of resources to the user, such as computation and storage. By designing different strategy graphs, our approach can solve the problem not only in the single-resource case but also in the multi-resource case.

We show our general system model in Fig. 4.6. The mobile server, which is a small-box data center consisting of several multi-core computers [57], are considered as the servers in the user-dense MEC systems. They are working independently without a central controller and are denoted as $\mathscr{S} = \{1, 2, \cdots, s\}$. Furthermore, those servers may provide either single resource or multiple different resources to the mobile users. On the other hand, users denoted by $\mathscr{U} = \{1, 2, \cdots, u\}$ are only interested in their own profit. The jobs of all users are assumed to be homogeneous and indistinguishable without priorities, deadlines, and multiple versions [67] because heterogeneous workload needs to be divided into different populations, which falls into the framework of multi-population mean field game [68, 69]. After defining server set $\mathscr{S}$ and user set $\mathscr{U}$, we can construct the system models in both single-resource case and multi-resource case in the following subsections. For more details regarding the computation of the solution to the MFG-based load-balancing problem, the reader may refer to [70].

4.4.1   Single-Resource Case

User i's jobs ($i \in \mathscr{U}$) are generated according to a Poisson process with mean rate λ_i, and $\hat{\lambda} = \sum_{i=1}^{u} \lambda_i$ denotes the total arrival rate of all users. User i will seek its own best strategy to allocate its jobs to all servers. Server j ($j \in \mathscr{S}$) is modeled by $M/M/1$ [71] with a service rate μ_j. To keep a stable system, we need to ensure that $\hat{\lambda} < \sum_{j=1}^{s} \mu_j$ and the system utilization rate is $\theta = \frac{\hat{\lambda}}{\sum_{j=1}^{s} \mu_j}$.

User i's strategy is denoted as $\chi_i \in \mathbb{R}^s$. χ_{ij} represents the proportion of user i's total amount of jobs assigned to server j. Therefore, user i's strategy can be considered as a mixed strategy on the server set $\mathscr{S}$. We record all users' strategies in one strategy profile $\mathscr{X} = (\chi_i)_{i=1}^{u}$, $\mathscr{X} \in \mathbb{R}^{u \times s}$ and each row χ_i is the mixed strategy of user i. The distribution of users' jobs on the servers is the "mean field".

When users determine their strategies, they want to know their individual response time. In order to obtain the expected response time of user i, we should first know the response time of server j to user i because user i's jobs are assigned on different servers. Under the single-resource case, the response time of server j is given by

$$R_j(\mathscr{X}) = \frac{1}{\mu_j - \sum_{n=1}^{u} \lambda_n \chi_{nj}}, \tag{4.31}$$

where μ_j is server j's service rate and $\sum_{n=1}^{u} \lambda_n \chi_{nj}$ is total workload on server j. Accordingly, the total expected response time is given by

$$T_i(\mathscr{X}) = \sum_{j=1}^{s} \chi_{ij} R_j(\mathscr{X}) = \sum_{j=1}^{s} \frac{\chi_{ij}}{\mu_j - \sum_{n=1}^{u} \lambda_n \chi_{nj}}, \tag{4.32}$$

where $\chi_{ij} R_j(\mathscr{X})$ is the response time of server j to user i weighted by user i's probability of selecting server j.

When we try to find the optimal strategies for users to minimize their expected response time, we need to satisfy the following constraints according to [72–75]

$$\chi_{ij} \geqslant 0, \forall i \in \mathscr{U}, \forall j \in \mathscr{S}, \tag{4.33}$$

$$\sum_{j=1}^{s} \chi_{ij} = 1, \forall i \in \mathscr{U}, \tag{4.34}$$

$$\sum_{n=1}^{u} \lambda_n \chi_{nj} \leqslant \mu_j, \forall j \in \mathscr{S}. \tag{4.35}$$

Equation (4.33) is recommended because χ_{ij} is the percentage of user i's jobs assigned to server j. Meanwhile, Eq. (4.34) needs to be satisfied because user i will allocate his jobs on all or some of the servers. In (4.35), $\sum_{n=1}^{u} \lambda_n \chi_{nj}$ is the workload on server j. Therefore, we need to keep the workload less than the service rate μ_j to ensure the stability of the system.

We regard the Nash Equilibrium [76], where all users have no incentive to modify their strategies unilaterally, as the solution to this non-cooperative population game. When we seek the optimal strategies to achieve the Nash Equilibrium, we will update the strategy of user i while keeping strategies of other users fixed with our mean field evolutionary methodology. Considering this, solving the non-cooperative population game for multi-users is equivalent to solving the following optimization problem for a single user

$$\min_{\mathscr{X}} \ T_i(\mathscr{X}) = \sum_{j=1}^{s} \frac{\chi_{ij}}{\mu_j - \sum_{n=1}^{u} \lambda_n \chi_{nj}}, \tag{4.36}$$

$$\text{s.t. } \chi_{ij} \geq 0, \forall j \in \mathscr{S}, \ \sum_{j=1}^{s} \chi_{ij} = 1, \ \sum_{n=1}^{u} \lambda_n \chi_{nj} < \mu_j, \forall j \in \mathscr{S},$$

where $T_i(\mathscr{X})$ is defined in (4.32) and the three constraints are (4.33), (4.34), and (4.35), respectively.

4.4.2 Multi-Resource Case

For the multi-resource case, each server provides multiple different resources for users to complete corresponding types of jobs. Therefore, aside from user set $\mathscr{U}$ and server set $\mathscr{S}$, we need to define the resource set as $\mathscr{R} = \{1, 2, \cdots, r\}$.

User i is generating the jobs of class i requiring r different resources according to a Poisson process with mean rate $\lambda_i = (\lambda_{i1}, \lambda_{i2}, \cdots, \lambda_{ir}), \lambda_i \in \mathbb{R}^r$. The kth entry of λ_i represents the rate of generating the kth type of jobs requiring the kth type of resources. The job profile $\lambda = (\lambda_i)_{i=1}^{u}$ records all users' job generating rates. These jobs are assigned to all the servers and each server is modeled by $M/M/1$ [71]. The service rate of server j is $\mu_j = (\mu_{jk})_{k=1}^{r} \in \mathbb{R}^k$, with the kth entry denoting service rate of the kth type of jobs. Thus $\mu = (\mu_j)_{j=1}^{s}$ is the $s \times r$ matrix recording all servers' service rate.

Similar to the single-resource case, each user will allocate its different types of jobs to the servers. However, under the multi-resource case, a generic user i not only needs to determine which server to connect but also which type of resource to utilize. Therefore, the strategy set for multi-resource case is the $r \times s$ different combinations of server set $\mathscr{S}$ and resource set $\mathscr{R}$, which is defined as follows: $\hat{\mathscr{S}} = \{(j, k) : j \in \mathscr{S}, k \in \mathscr{R}\}$. Denoting the strategy of user i as $\pi_i \in \mathbb{R}^t, t = r \times s$, it is the probability distribution on $\hat{\mathscr{S}}$. The $j + s(k-1)$ entry represents the probability of selecting k type of resource on server j. π_i is recorded in this way because it is convenient for us to update π_i later with gradient matrix and our strategy graph is a 2-D lattice. Hence, $\pi = (\pi_i)_{i=1}^{u}$ is the strategy profile recording all users' strategies.

As user i's jobs are assigned to all the servers, we need to compute the response time of each server j in order to obtain user i's total response time. Moreover, the response time of server j should be the total response time of all types of jobs. Therefore, the derived response time of server j to user i will be

$$R_j(\pi) = \sum_{k=1}^{r} \frac{1}{\mu_{jk} - \sum_{n=1}^{u} \lambda_{nk}\pi_{nm}}, \tag{4.37}$$

where $m = j + s(k-1)$ with $j \in \mathscr{S}, k \in \mathscr{R}$ and $\sum_{n=1}^{u} \lambda_{nk}\pi_{nm}$ is the total kth type of jobs assigned on server j. Consequently, the total expected response time of user i is computed by

$$T_i(\pi) = \sum_{j=1}^{s} \pi_{im}R_j(\pi) = \sum_{j=1}^{s}\sum_{k=1}^{r} \frac{\pi_{im}}{\mu_{jk} - \sum_{n=1}^{u} \lambda_{nk}\pi_{nm}}, \tag{4.38}$$

where $\pi_{im}R_j(\pi)$ is server j's response time weighted by the probability of selecting resource k on server j.

Similar to the single-resource case, certain preferences are recommended by [72–75]. They are given as follows:

$$\pi_{im} \geqslant 0, \tag{4.39}$$

$$\sum_{j=1}^{s}\sum_{k=1}^{r} \pi_{im} = 1, \tag{4.40}$$

$$\sum_{n=1}^{u} \lambda_{nk}\pi_{nm} < \mu_{jk}, \tag{4.41}$$

where $m = j + s(k-1), i \in \mathscr{U}, j \in \mathscr{S}, k \in \mathscr{R}$. The first and second are suggested because π_{im} is probability distribution on $\hat{\mathscr{S}}$. The third needs to be satisfied for stability of the whole system.

With the system constraints defined above and the expected response time $T_i(\pi)$ as the objective function, we still formulate our problem for the multi-resource case as the following optimization problem for a generic user i:

$$\min_{\pi} \sum_{j=1}^{s}\sum_{k=1}^{r} \frac{\pi_{im}}{\mu_{jk} - \sum_{n=1}^{u} \lambda_{nk}\pi_{nm}}, \tag{4.42}$$

$$\text{s.t. } \pi_{im} \geqslant 0, \ \sum_{j=1}^{s}\sum_{k=1}^{r} \pi_{im} = 1, \ \sum_{n=1}^{u} \lambda_{nk}\pi_{nm} < \mu_{jk},$$

where $m = j + s(k-1), j \in \mathscr{S}, k \in \mathscr{R}$. The objective function in (4.42) is the total expected response time defined in (4.38) and the three constraints are (4.39), (4.40), and (4.41), respectively.

For more details regarding the computation of the solution to the MFG-based load-balancing problems for both cases, the reader may refer to [70].

References

1. L. Song, D. Niyato, Z. Han, E. Hossain, *Wireless Device-to-Device Communications and Networks* (Cambridge University Press, Cambridge, 2015)
2. N. Bhushan, J. Li, D. Malladi, R. Gilmore, D. Brenner, A. Damnjanovic, R.T. Sukhavasi, C. Patel, S. Geirhofer, Network densification: the dominant theme for wireless evolution into 5G. IEEE Commun. Mag. **52**(2), 82–89 (2014)
3. B. Soret, K.I. Pedersen, N.T. Jørgensen, V. Fernandez-Lopez, Interference coordination for dense wireless networks. IEEE Commun. Mag. **53**(1), 102–109 (2015)
4. H. Zhang, C. Jiang, J. Cheng, V.C. Leung, Cooperative interference mitigation and handover management for heterogeneous cloud small cellnetworks. IEEE Wirel. Commun. **22**(3), 92–99 (2015)
5. C. Yang, J. Li, M. Guizani, Cooperation for spectral and energy efficiency in ultra-dense small cell networks. IEEE Wirel. Commun. **23**(1), 64–71 (2016)
6. F. Wang, C. Xu, L. Song, Z. Han, Energy-efficient resource allocation for device-to-device underlay communication. IEEE Trans. Wirel. Commun. **14**(4), 2082–2092 (2014)
7. R. Yin, C. Zhong, G. Yu, Z. Zhang, K.K. Wong, X. Chen, Joint spectrum and power allocation for D2D communications underlaying cellular networks. IEEE Trans. Veh. Technol. **65**(4), 2182–2195 (2015)
8. C.-H. Yu, O. Tirkkonen, K. Doppler, C. Ribeiro, Power optimization of device-to-device communication underlaying cellular communication, in *2009 IEEE International Conference on Communications* (IEEE, Piscataway, 2009), pp. 1–5
9. P. Janis, V. Koivunen, C. Ribeiro, J. Korhonen, K. Doppler, K. Hugl, Interference-aware resource allocation for device-to-device radio underlaying cellular networks, in *VTC Spring 2009-IEEE 69th Vehicular Technology Conference* (IEEE, Piscataway, 2009), pp. 1–5
10. G. Fodor, N. Reider, A distributed power control scheme for cellular network assisted D2D communications, in *2011 IEEE Global Telecommunications Conference-GLOBECOM 2011*, (IEEE, Piscataway, 2011), pp. 1–6
11. S. Maghsudi, S. Stanczak, Hybrid centralized–distributed resource allocation for device-to-device communication underlaying cellular networks. IEEE Trans. Veh. Technol. **65**(4), 2481–2495 (2015)
12. K. Zhu, E. Hossain, Joint mode selection and spectrum partitioning for device-to-device communication: a dynamic Stackelberg game. IEEE Trans. Wirel. Commun. **14**(3), 1406–1420 (2014)
13. J. Huang, Y. Zhao, K. Sohraby, Resource allocation for intercell device-to-device communication underlaying cellular network: a game-theoretic approach, in *2014 23rd International Conference on Computer Communication and Networks (ICCCN)* (IEEE, Piscataway, 2014), pp. 1–8
14. R. Yin, G. Yu, H. Zhang, Z. Zhang, G. Y. Li, Pricing-based interference coordination for D2D communications in cellular networks. IEEE Trans. Wirel. Commun. **14**(3), 1519–1532 (2014)
15. L. Song, D. Niyato, Z. Han, E. Hossain, Game-theoretic resource allocation methods for device-to-device communication. IEEE Wirel. Commun. **21**(3), 136–144 (2014)
16. Y. Shen, C. Jiang, T.Q. Quek, H. Zhang, Y. Ren, Device-to-device cluster assisted downlink video sharing—a base station energy saving approach, in *2014 IEEE Global Conference on Signal and Information Processing (GlobalSIP)* (IEEE, Piscataway, 2014), pp. 108–112
17. S.M. Azimi, M.H. Manshaei, F. Hendessi, Hybrid cellular and device-to-device communication power control: Nash bargaining game, in *7th International Symposium on Telecommunications (IST'2014)* (IEEE, Piscataway, 2014), pp. 1077–1081
18. Y. Xiao, K.-C. Chen, C. Yuen, Z. Han, L.A. DaSilva, A bayesian overlapping coalition formation game for device-to-device spectrum sharing in cellular networks. IEEE Trans. Wirel. Commun. **14**(7), 4034–4051 (2015)

19. X. Lu, P. Wang, D. Niyato, Hierarchical cooperation for operator-controlled device-to-device communications: a layered coalitional game approach, in *2015 IEEE Wireless Communications and Networking Conference (WCNC)* (IEEE, Piscataway, 2015), pp. 2056–2061
20. J.-M. Lasry, P.-L. Lions, Mean field games. Jpn. J. Math. **2**(1), 229–260 (2007)
21. O. Gueant, J.-M. Lasry, P.-L. Lions, Mean field games and applications, in *Paris-Princeton Lectures on Mathematical Finance 2010* (Springer, Berlin, 2011), pp. 205–266
22. Z. Han, D. Niyato, W. Saad, T. Basar, A. Hjørungnes, *Game Theory in Wireless and Communication Networks: Theory, Models, and Applications* (Cambridge University Press, Cambridge, 2012)
23. H. Tembine, R. Tempone, P. Vilanova, Mean field games for cognitive radio networks, in *2012 American Control Conference (ACC)* (IEEE, Piscataway, 2012), pp. 6388–6393
24. F. Meriaux, S. Lasaulce, Mean-field games and green power control, in *International Conference on Network Games, Control and Optimization (NetGCooP 2011)* (IEEE, Piscataway, 2011), pp. 1–5
25. A.Y. Al-Zahrani, F.R. Yu, M. Huang, A joint cross-layer and colayer interference management scheme in hyperdense heterogeneous networks using mean-field game theory. IEEE Trans. Veh. Technol. **65**(3), 1522–1535 (2015)
26. Y. Zhang, C. Yang, J. Li, Z. Han, Distributed interference-aware traffic offloading and power control in ultra-dense networks: mean field game with dominating player. IEEE Trans. Veh. Technol. **68**(9), 8814–8826 (2019)
27. P. Semasinghe, E. Hossain, Downlink power control in self-organizing dense small cells underlaying macrocells: a mean field game. IEEE Trans. Mob. Comput. **15**(2), 350–363 (2015)
28. A.F. Hanif, H. Tembine, M. Assaad, D. Zeghlache, Mean-field games for resource sharing in cloud-based networks. IEEE/ACM Trans. Netw. **24**(1), 624–637 (2015)
29. R. Couillet, S.M. Perlaza, H. Tembine, M. Debbah, Electrical vehicles in the smart grid: a mean field game analysis. IEEE J. Sel. Areas Commun. **30**(6), 1086–1096 (2012)
30. Y. Zhang, C. Yang, J. Li, Z. Han, Distributed interference-aware traffic offloading and power control in ultra-dense networks: mean field game with dominating player. IEEE Trans. Veh. Technol. **68**(9), 8814–8826 (2019)
31. X. Ge, S. Tu, G. Mao, C.-X. Wang, T. Han, 5G ultra-dense cellular networks. IEEE Wirel. Commun. **23**(1), 72–79 (2016)
32. J.G. Andrews, S. Buzzi, W. Choi, S.V. Hanly, A. Lozano, A.C. Soong, J.C. Zhang, What will 5G be? IEEE J. Sel. Areas Commun. **32**(6), 1065–1082 (2014)
33. M. Kamel, W. Hamouda, A. Youssef, Ultra-dense networks: a survey. IEEE Commun. Surv. Tutorials **18**(4), 2522–2545 (2016)
34. F.A. Andrade, R. Storvold, T.A. Johansen, Autonomous UAV surveillance of a ship's path with MPC for maritime situational awareness, in *2017 International Conference on Unmanned Aircraft Systems (ICUAS)* (IEEE, Piscataway, 2017), pp. 633–639
35. H. Shakhatreh, A.H. Sawalmeh, A. Al-Fuqaha, Z. Dou, E. Almaita, I. Khalil, N.S. Othman, A. Khreishah, M. Guizani, Unmanned aerial vehicles (UAVs): a survey on civil applications and key research challenges. IEEE Access **7**, 48572–48634 (2019)
36. V. Sharma, M. Bennis, R. Kumar, UAV-assisted heterogeneous net-works for capacity enhancement. IEEE Commun. Lett. **20**(6), 1207–1210 (2016)
37. L. Li, Z. Zhang, K. Xue, M. Wang, M. Pan, Z. Han, AI-aided downlink interference control in dense interference-aware drone small cells networks. IEEE Access **8**, 15110–15122 (2020)
38. J.-M. Lasry, P.-L. Lions, Mean field games. Jpn. J. Math. **2**(1), 229–260 (2007)
39. Y. Achdou, F. Camilli, I. Capuzzo-Dolcetta, Mean field games: numerical methods for the planning problem. SIAM J. Control Optim. **50**(1), 77–109 (2012)
40. Q. Cheng, L. Li, Y. Sun, D. Wang, W. Liang, X. Li, Z. Han, Efficient resource allocation for NOMA-MEC system in ultra-dense network: a mean field game approach, in *2020 IEEE International Conference on Communications Workshops (ICC Workshops)* (IEEE, Piscataway, 2020), pp. 1–6

41. Y. Sun, L. Li, Q. Cheng, D. Wang, W. Liang, X. Li, Z. Han, Joint trajectory and power optimization in multi-type UAVs network with mean field Q-learning, in *2020 IEEE International Conference on Communications Workshops (ICC Workshops)* (IEEE, Piscataway, 2020), pp. 1–6
42. R.S. Sutton, A.G. Barto, *Reinforcement Learning: An Introduction* (MIT Press, Cambridge, 2018)
43. H. Gacanin, Autonomous wireless systems with artificial intelligence: a knowledge management perspective. IEEE Veh. Technol. Mag. **14**(3), 51–59 (2019)
44. F. Tang, Y. Kawamoto, N. Kato, J. Liu, Future intelligent and secure vehicular network toward 6G: Machine-learning approaches. Proc. IEEE **108**(2), 292–307 (2019)
45. F. Tang, Z.M. Fadlullah, B. Mao, N. Kato, An intelligent traffic load prediction-based adaptive channel assignment algorithm in SDN-IoT: a deep learning approach. IEEE Internet Things J. **5**(6), 5141–5154 (2018)
46. N. Kato, B. Mao, F. Tang, Y. Kawamoto, J. Liu, Ten challenges in advancing machine learning technologies toward 6G. IEEE Wirel. Commun. **27**(3), 96–103 (2020)
47. Y. Wang, J. Yang, M. Liu, G. Gui, LightAMC: lightweight automatic modulation classification via deep learning and compressive sensing. IEEE Trans. Veh. Technol. **69**(3), 3491–3495 (2020)
48. G. Gui, F. Liu, J. Sun, J. Yang, Z. Zhou, D. Zhao, Flight delay prediction based on aviation big data and machine learning. IEEE Trans. Veh. Technol. **69**(1), 140–150 (2019)
49. F. Tang, B. Mao, Z.M. Fadlullah, N. Kato, On a novel deep-learning-based intelligent partially overlapping channel assignment in SDN-IoT. IEEE Commun. Mag. **56**(9), 80–86 (2018)
50. N. Kato, Z.M. Fadlullah, F. Tang, B. Mao, S. Tani, A. Okamura, J. Liu, Optimizing space-air-ground integrated networks by artificial intelligence. IEEE Wirel. Commun. **26**(4), 140–147 (2019)
51. M. Chen, U. Challita, W. Saad, C. Yin, M. Debbah, Artificial neural networks-based machine learning for wireless networks: a tutorial. IEEE Commun. Surv. Tutorials **21**(4), 3039–3071 (2019)
52. V. Mnih, K. Kavukcuoglu, D. Silver, A.A. Rusu, J. Veness, M.G. Belle-mare, A. Graves, M. Riedmiller, A.K. Fidjeland, G. Ostrovski et al., Human-level control through deep reinforcement learning. Nature **518**(7540), 529–533 (2015)
53. M. Mozaffari, W. Saad, M. Bennis, M. Debbah, Mobile unmanned aerial vehicles (UAVs) for energy-efficient internet of things communications. IEEE Trans. Wirel. Commun. **16**(11), 7574–7589 (2017)
54. A. Farajzadeh, O. Ercetin, H. Yanikomeroglu, UAV data collection over NOMA backscatter networks: UAV altitude and trajectory optimization, in *ICC 2019-2019 IEEE International Conference on Communications (ICC)* (IEEE, Piscataway, 2019), pp. 1–7
55. F. Shams, G. Bacci, M. Luise, A Q-learning game-theory-based algorithm to improve the energy efficiency of a multiple relay-aided network, in *2014 XXXIth URSI General Assembly and Scientific Symposium (URSIGASS)* (IEEE, Piscataway, 2014), pp. 1–4
56. L. Li, Q. Cheng, K. Xue, C. Yang, Z. Han, Downlink transmit power control in ultra-dense UAV network based on mean field game and deep reinforcement learning. IEEE Trans. Veh. Technol. **69**(12), 15594–15605 (2020)
57. N. Abbas, Y. Zhang, A. Taherkordi, T. Skeie, Mobile edge computing: a survey. IEEE Internet Things J. **5**(1), 450–465 (2018)
58. Cisco visual networking index: Global mobile data traffic forecast update, 2017-2022 white paper, Feb 2019. https://www.cisco.com/c/en/us/solutions/collateral/service-provider/visual-networking-index-vni/white-paper-c11-738429.html
59. Z. Han, D. Niyato, W. Saad, T. Basar, *Game Theory for Next Generation Wireless and Communication Networks: Modeling, Analysis, and Design* (Cambridge University Press, Cambridge, 2019)
60. D. T. Hoang, D. Niyato, D.N. Nguyen, E. Dutkiewicz, P. Wang, Z. Han, A dynamic edge caching framework for mobile 5G networks. IEEE Wirel. Commun. **25**(5), 95–103 (2018)

61. Y. Zhang, C. Yang, J. Li, Z. Han, Distributed interference-aware traffic offloading and power control in ultra-dense networks: mean field game with dominating player. IEEE Trans. Veh. Technol. **68**(9), 8814–8826 (2019)
62. Y. Teng, M. Liu, F.R. Yu, V.C.M. Leung, M. Song, Y. Zhang, Resource allocation for ultra-dense networks: a survey, some research issues and challenges. IEEE Commun. Surv. Tutorials **21**(3), 2134–2168 (2019)
63. C. Yang, J. Li, M. Sheng, A. Anpalagan, J. Xiao, Mean field game-theoretic framework for interference and energy-aware control in 5g ultra-dense networks. IEEE Wirel. Commun. **25**(1), 114–121 (2018)
64. Z. Han, D. Niyato, W. Saad, T. Basar, A. Hjørungnes, *Game Theory in Wireless and Communication Networks: Theory, Models, and Applications* (Cambridge University Press, Cambridge, 2012)
65. R.A. Banez, L. Li, C. Yang, L. Song, Z. Han, A mean-field-type game approach to computation offloading in mobile edge computing networks, in *Proceedings of the 2019 IEEE International Conference on Communications (ICC), Shanghai*, May 2019, pp. 1–6
66. T.Q. Duong, X. Chu, H.A. Suraweera, *Mean Field Games for 5G Ultra-dense Networks: A Resource Management Perspective* (Wiley, Hoboken, 2019), pp. 65–89
67. T.D. Braun, H.J. Siegel, A.A. Maciejewski, Y. Hong, Static resource allocation for heterogeneous computing environments with tasks having dependencies, priorities, deadlines, and multiple versions. J. Parallel Distrib. Comput. **68**(11), 1504–1516 (2008)
68. R.A. Banez, H. Gao, L. Li, C. Yang, Z. Han, H.V. Poor, Belief and opinion evolution in social networks based on a multi-population mean field game approach, in *Proceedings of the 2020 IEEE International Conference on Communications, Dublin*, Jun 2020, pp. 1–6
69. A. Bensoussan, T. Huang, M. Lauriere, Mean field control and mean field game models with several populations. Optim. Control **3**, 173–209 (2018)
70. H. Gao, W. Li, R.A. Banez, Z. Han, H.V. Poor, Mean field evolutionary dynamics in dense-user multi-access edge computing systems. IEEE Trans. Wirel. Commun. **19**(12), 7825–7835 (2020)
71. K.S. Trivedi, *Probability and Statistics with Reliability, Queuing, and Computer Science Applications* (Wiley, Chichester, 2016)
72. D. Grosu, A.T. Chronopoulos, Noncooperative load balancing in distributed systems. J. Parallel Distrib. Comput. **65**(9), 1022–1034 (2005)
73. S. Penmatsa, A.T. Chronopoulos, Game-theoretic static load balancing for distributed systems. J. Parallel Distrib. Comput. **71**(4), 537–555 (2011)
74. H. Kameda, *Optimal Load Balancing in Distributed Computer Systems* (Springer, Berlin, 1997)
75. X. Tang, S.T. Chanson, Optimizing static job scheduling in a network of heterogeneous computers, in *Proceedings 2000 International Conference on Parallel Processing, Toronto*, Aug 2000, pp. 373–382
76. S.-N. Chow, W. Li, J. Lu, H. Zhou, Equilibrium selection via optimal transport. arXiv:1703.08442 [math] (2017)

Chapter 5
Multiple-Population Mean Field Game for Social Networks

The number of users engaged in social media through social networks continues to grow as people become more passionate on current social topics and issues. When social network users share similar characteristics such as political orientation, age, and gender, they can be grouped and analyzed according to these similarities. Meanwhile, the behavior of users in a social network setting can be extracted from their belief and opinions regarding social topics and issues. Furthermore, understanding user behavior has been a priority of network service providers. Integrating the idea of user behavior modeling to a multiple-group social network, a multiple-population mean field game (MPMFG) approach is proposed to analyze and model the belief and opinion evolution of users in a social network. Through the proposed model, information can be gained on the behavior of social network users belonging to different groups. Specifically, the proposed model can be utilized to estimate and predict how a social network group affects the belief and opinion of other groups. Simulations are provided to show the belief and opinion evolution of users in a multiple-population setting. Theoretical results and experimental results using a social evolution dataset are presented to demonstrate the effectiveness of the proposed MFG approach.

5.1 System Model

In this section, the opinion dynamics equation that characterizes the change or update of opinion of a user in a single-population social network is derived. By single population, it is meant that the users share a common characteristic (e.g., age, gender, and political orientation). Then, the cost function that characterizes a user's behavior or preferences in a social network is described.

In the following derivations, consider a social network with N users denoted by graph $\mathscr{G} = (\mathscr{V}, \mathscr{E})$, where $\mathscr{V}$ is the set of all social network users and $\mathscr{E} \subseteq \mathscr{V} \times \mathscr{V}$

© The Editor(s) (if applicable) and The Author(s), under exclusive license to
Springer Nature Switzerland AG 2021
R. A. Banez et al., *Mean Field Game and its Applications in Wireless Networks*,
Wireless Networks, https://doi.org/10.1007/978-3-030-86905-2_5

is the set of all links between two distinct users in $\mathcal{V}$. Two distinct users that are linked are neighbors, and hence the neighborhood of a user refers to the set of all its neighbors. Also, denote x_i, $i = 1, \ldots, N$ as the state or scalar opinion of user i in a social network.

5.1.1   Opinion Dynamics Equation

The state or opinion $x_i \in \mathbb{R}$ of user i evolves with time k depending on the opinions of its neighbors. In this work, the following opinion dynamics equation is adapted where at each stage $k = 0, 1, \ldots$, the opinion of user i evolves according to

$$x_i(k + 1) = \frac{\sum_{j=1}^{N} \alpha_{ij} x_j(k)}{\sum_{j=1}^{N} \alpha_{ij}} + \beta_i u_i(x_i, k). \qquad (5.1)$$

The first term refers to the Hegselmann-Krause (HK) model of opinion evolution [1] where α_{ij} denotes the adjacency between users i and j, with $\alpha_{ij} = 1$ if users i and j are connected and $\alpha_{ij} = 0$ if they are not connected. Therefore, α_{ij} determines the neighbors of user i. The second term has been added to reflect the change in opinion based on the control effort $u_i(x_i, k)$ with β_i as a constant [2]. Figure 5.1a shows the system model for a single-population social network. The blue-colored users belong to the neighborhood of user i while the gray-colored users are outside its neighborhood.

Social network users can be stubborn which means they consider their initial belief or opinion together with the belief or opinion of their neighbors. A popular social network dynamic model that considers stubborn agents is the FJ model [3]. In the FJ model, the influence of neighbor j to user i is quantified by the interpersonal influence w_{ij}, which refers to the strength of influence of between social network users i and j. Moreover, the willingness of user i to external influence is denoted by λ_i, which also refers to the likeliness of user i to change from its prejudice or initial opinion x_{i0}. The evolution of opinion of user i according to the FJ model is given by

$$x_i(k + 1) = \lambda_i \sum_{j=1}^{N} w_{ij} x_j(k) + (1 - \lambda_i) x_{i0}. \qquad (5.2)$$

In this work, the HK model in (5.1) is integrated to (5.2) where the opinion of user i depends on the average of the opinion of its neighbors. Thus, the opinion dynamics equation for social networks with stubborn agents in (5.1) is equivalent to

$$x_i(k + 1) = \lambda_i \frac{\sum_{j=1}^{N} \alpha_{ij} x_j(k)}{\sum_{j=1}^{N} \alpha_{ij}} + (1 - \lambda_i) x_{i0} + \beta_i u_i(x_i, k), \qquad (5.3)$$

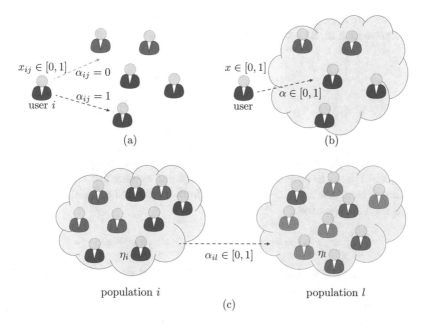

(a)

(b)

population i population l

(c)

Fig. 5.1 Illustration of a social network with single population and multiple populations

where the interpersonal influence w_{ij} in (5.2) has been expressed as $w_{ij} = \alpha_{ij}/\sum \alpha_{ij}$. Note that if $\lambda_i = 1$, the stubborn case in (5.3) reduces to the non-stubborn case in (5.1).

In continuous time, the corresponding opinion dynamics equation for social networks is

$$dx_i(t) = \left(\lambda_i \frac{\sum_{j=1}^{N} \alpha_{ij} x_j(t)}{\sum_{j=1}^{N} \alpha_{ij}} + (1 - \lambda_i)x_{i0} - x_i(t) + \beta_i u_i(x_i, t) \right) dt$$

$$= f_i(x_i, x_{-i}, u_i) \, dt, \tag{5.4}$$

where λ_i denotes the degree of non-stubbornness, $x_{-i} = (x_1, \ldots, x_{i-1}, x_{i+1}, \ldots, x_n)$ refers to the state profile containing the states of users other than user i.

5.1.2 Cost Function

Let $u_i \in \mathcal{U}_i$ be the control or influence effort of a user in population i with opinion x_i at time t. In a social network, a user can influence anybody in the network through various ways such as friendship links, advertisements, and social network posts.

By controlling u_i, user i can minimize its own local cost function r_i that refers to the accumulation of the running cost function r_i from $t = 0$ to T,

$$J_i(u_i, u_{-i}) = \int_0^T r_i(x_i, u_i, u_{-i}) \, dt, \qquad (5.5)$$

where $u_{-i} = (u_1, \ldots, u_{i-1}, u_{i+1}, \ldots, u_n)$ denotes the control profile of all users except user i.

The running cost $r_i(x_i, u_i, u_{-i})$ consists of the following: the control effort cost associated with the influence u_i; the popularity cost corresponding to the cost or reward when user i has different or the same opinion as other users; and the opinion cost that depends on how far opinion x_i is from the desired or ideal opinion x_d of the population. Therefore,

$$r_i(x_i, u_i, u_{-i}) = c_1 u_i^2 + c_2 \mu(x_i, x_{-i}) + c_3 \|x_i - x_d\|, \qquad (5.6)$$

where c_1, c_2, and c_3 are constants, and $\mu(x_i, x_{-i})$ is the relative frequency distribution, $\mu(x_i, x_{-i}) = \frac{1}{N} \sum_{j=1}^N \delta_{x_j = x_i}$, with $\delta = 1$ if $x_j = x_i$ and $\delta = 0$ if $x_j \neq x_i$. Note that the state x_i of user i depends on its control u_i through (5.3). Consequently, the cost function r_i depends on the control profile (u_i, u_{-i}) through $\mu(x_i, x_{-i})$, and it can be written as $r_i(x_i, u_i, u_{-i})$.

When stubborn agents are considered, the cost function will include a trade-off between the distance of opinion x_i to the ideal or target opinion x_d and to the initial opinion x_{i0}. Thus,

$$r_i(x_i, u_i, u_{-i}) = c_1 u_i^2 + c_2 \mu(x_i, x_{-i}) + c_3 \lambda_i \|x_i - x_d\| + c_4 (1 - \lambda_i) \|x_i - x_{i0}\|, \qquad (5.7)$$

where c_4 is a constant. Note that if $\lambda_i = 1$, the stubborn case in (5.7) reduces to the non-stubborn case in (5.6).

The goal of each social network user i is to vary the control u_i to minimize the cost $J_i(u_i, u_{-i})$ subject to the opinion dynamics $dx_i(t)$,

$$\min_{u_i \in \mathcal{U}_i} \quad J_i(u_i, u_{-i}) = \int_0^T r_i(x_i, u_i, u_{-i}) \, dt,$$
$$\text{subject to} \quad dx_i(t) = f_i(x_i, u_i, u_{-i}) \, dt. \qquad (5.8)$$

In the following section, the concepts and motivation behind MFGs are introduced. Then, the opinion dynamics and cost functions are reformulated using an MFG framework.

5.2 Mean Field Game with Single Population

5.2.1 Background

The calculation of the Nash equilibrium of a differential game with N players involves solving N coupled HJB equations. It becomes more complicated as the number of players increases because of increased interactions and coupling between the players. As a consequence, mean field games (MFGs) have been proposed to reformulate the game problem. MFGs were introduced by Lasry and Lions in [4] and have been applied in many applications in economics and engineering. MFGs can be utilized when the number of players is large and when the players are indistinguishable yet can have heterogeneous states. In an MFG, the aggregate effect of all the players is considered rather than the individual effect of each player.

Since the number of players in an MFG is large and the players are indistinguishable, the game can be seen as a representative or reference player playing against the mean field or aggregate behavior of other players. An MFG can be characterized by a pair of partial differential equations (PDEs): an HJB equation that corresponds to the evolution of the value function (i.e., optimized objective) of a player in response to the mean field, and a Fokker-Planck-Kolmogorov (FPK) equation that corresponds to the evolution of the mean field of players that are behaving optimally.

Consider a representative player with state $x \in \mathcal{X}$ and control $u \in \mathcal{U}$. Suppose m is the mean field or the distribution of the state of the players. An MFG can be expressed as a pair of HJB and FPK equations,

$$
\begin{aligned}
-\frac{\partial v}{\partial t}(x, t) - H(x, m, p, t) &= \frac{\sigma^2}{2} \Delta_x v(x, t), \\
\frac{\partial m}{\partial t}(x, t) + \operatorname{div}\big[f(x, u, m, t)m(x, t)\big] &= \frac{\sigma^2}{2} \Delta_x m(x, t),
\end{aligned}
\tag{5.9}
$$

where Δ_x refers to the Laplace operator and div refers to the divergence operator. The function $H(x, m, p, t)$ is called the Hamiltonian, and it is defined mathematically as

$$
H(x, m, p, t) = \min_{u \in \mathcal{U}} \big[r(x, u, m, t) + f(x, u, m, t) \cdot p(t)\big],
\tag{5.10}
$$

with boundary conditions $v(x, T) = g(x, m(x, T))$ and $m(x, 0) = m_0(x)$, where $p(t) = \nabla_x v(x, t)$. The first equation in (5.9) is the HJB equation that characterizes the optimized reaction of a player with the mean field, while the second equation in (5.9) is the FPK equation that describes the evolution of the population that behaves optimally [5]. Note that the profile control u_{-i} found in differential games has now been replaced by mean field m in MFG to imply the dependence of the game on the collective behavior of the players rather than their individual influence.

The function $m(x, t)$ refers to the mean field, and it corresponds to the distribution of the states of the players with respect to time. The formal definition of $m(x, t)$ is given as follows.

Definition 5.1 The mean field $m(x, t)$ denotes the probability distribution of the players with state x at time t. Mathematically,

$$m(x, t) = \lim_{N \to \infty} \frac{1}{N} \left(\sum_{i=1}^{N} \delta_{x_i = x} \right), \tag{5.11}$$

where $\delta = 1$ if $x_i = x$ and $\delta = 0$ if $x_i \neq x$.

The following theorem defines an MFG problem and provides insightful proof of the HJB and FPK equations. For more comprehensive analysis regarding MFGs, the reader may refer to [4].

Theorem 5.1 *Consider a noncooperative game among large number of indistinguishable players. If every player faces the optimization problem*

$$\min_{u \in \mathcal{U}} \quad J(u, m) = \mathbb{E}\left[\int_0^T r(x, u, m, t)\, dt + g(x(T), m(x, T)) \right],$$

$$\text{subject to} \quad dx(t) = f(x, u, m, t)\, dt + \sigma\, dw(t), \tag{5.12}$$

$$x(0) = x_0,$$

the equivalent noncooperative mean field game is represented by the pair of HJB and FPK equations,

$$-\frac{\partial v}{\partial t}(x, t) - H(x, m, p, t) = \frac{\sigma^2}{2} \Delta_x v(x, t),$$

$$\frac{\partial m}{\partial t}(x, t) + \text{div}\big[f(x, u, m, t) m(x, t) \big] = \frac{\sigma^2}{2} \Delta_x m(x, t), \tag{5.13}$$

where the boundary conditions are $v(x, T) = g(x, m(x, T))$ and $m(x, 0) = m_0(x)$, and the Hamiltonian $H(x, m, p, t) = \min_{u \in \mathcal{U}} \big[r(x, u, m, t) + f(x, u, m, t) \cdot p(t) \big]$, where $p(t) = \nabla_x v(x, t)$.

Proof Let the value function $v(x, t) = \min_{u \in \mathcal{U}} J(u, m)$. Applying the dynamic programming principle (DPP) for stochastic optimal control to $v(x, t)$ leads to

$$v(x, t) = \min_{u \in \mathcal{U}} \mathbb{E}\left[\int_t^{t+\delta} r(x, u, m, \tau)\, d\tau + v(x(t + \delta), t + \delta) \right],$$

where $0 < \delta < T$.

Using Ito stochastic differentiation rule, the term $v(x(t+\delta), t+\delta)$ can be written as

$$v(x(t+\delta), t+\delta) = v(x,t) + \int_t^{t+\delta} \left[\frac{\partial v}{\partial t}(x(\tau), \tau) + f(x, u, m, \tau) \cdot \nabla_x v(x(\tau), \tau) \right] d\tau$$
$$+ \int_t^{t+\delta} \frac{\sigma^2}{2} \Delta_x v(x(\tau), \tau) \, d\tau$$
$$+ \int_t^{t+\delta} \sigma \nabla_x v(x(\tau), \tau) \, dw(\tau).$$

Substituting this equation into the first equation of $v(x,t)$, taking the terms $v(x,t)$ and $\int_t^{t+\delta} \frac{\partial v}{\partial t}(x(\tau), \tau) \, d\tau$ outside of the min since they are independent of u, and cancelling $v(x,t)$ on both sides result in

$$\int_t^{t+\delta} \frac{\partial v}{\partial t}(x(\tau), \tau) \, d\tau$$
$$+ \min_{u \in \mathcal{U}} \mathbb{E} \left[\int_t^{t+\delta} \left[r(x, u, m, \tau) \, d\tau + f(x, u, m, \tau) \cdot \nabla_x v(x(\tau), \tau) \right] d\tau \right.$$
$$\left. + \int_t^{t+\delta} \frac{\sigma^2}{2} \Delta_x v(x(\tau), \tau) \, d\tau + \int_t^{t+\delta} \sigma \nabla_x v(x(\tau), \tau) \, dw(\tau) \right] = 0,$$

Dividing the result by δ, getting the limit as $\delta \to 0$, and evaluating the expectation lead to the HJB equation,

$$-\frac{\partial v}{\partial t}(x, t) - H(x, m, p, t) = \frac{\sigma^2}{2} \Delta_x v(x, t),$$

with $H(x, m, p, t) = \min_{u \in \mathcal{U}} \left[r(x, u, m, t) + f(x, u, m, t) \cdot p(t) \right]$ and $p(t) = \nabla_x v(x, t)$. This concludes the derivation of the HJB equation in (5.13).

The derivation of the FPK equation starts with letting ϕ be a twice-differentiable function of a stochastic process $x(t) = x$ with state dynamics $dx(t)$ governed by (5.12).

Apply Ito differentiation rule to get $d\phi(x)$,

$$d\phi(x) = \left[f(x, u, m, t) \cdot \nabla_x \phi(x) + \frac{\sigma^2}{2} \Delta_x \phi(x) \right] dt + \nabla_x \phi(x) \sigma \, dw(t).$$

Afterwards, divide by dt and take the expectation,

$$\mathbb{E} \left[\frac{d\phi(x)}{dt} \right] = \mathbb{E} \left[f(x, u, m, t) \cdot \nabla_x \phi(x) + \frac{\sigma^2}{2} \Delta_x \phi(x) \right],$$

since $\mathbb{E}[\nabla_x \phi(x) \sigma \, dw(t)] = 0$.

Rewriting in terms of the mean field $m(x, t)$ and using integration by parts lead to

$$\int_{\mathscr{X}} \frac{\partial \phi}{\partial x}(t) m(x, t)\, dx = \int_{\mathscr{X}} [f(x, u, m, t) \cdot \nabla_x \phi(x)] m(x, t)\, dx$$
$$+ \int_{\mathscr{X}} \frac{\sigma^2}{2} \Delta_x \phi(x) m(x, t)\, dx,$$

$$\int_{\mathscr{X}} \frac{\partial m}{\partial x}(x, t) \phi(x)\, dx = \int_{\mathscr{X}} -\mathrm{div}\big[f(x, u, m, t) m(x, t) \big] \phi(x, t)\, dx$$
$$+ \frac{\sigma^2}{2} \int_{\mathscr{X}} \Delta_x m(x, t) \phi(x)\, dx.$$

Differentiating the last equation with respect to x and dividing out $\phi(x)$ yield the FPK equation in (5.13),

$$\frac{\partial m}{\partial t}(x, t) + \mathrm{div}\big[f(x, u, m, t) m(x, t) \big] = \frac{\sigma^2}{2} \Delta_x m(x, t).$$

The following subsection is devoted to applying the single-population MFG to social network belief and opinion evolution. Using an MFG, the evolution with time of the opinions of the users in a huge social network can be characterized by a pair of HJB and FPK equations.

5.2.2   A Mean Field Game Problem for Single-Population Social Networks

Let $x \in \mathbb{R}$ be the scalar opinion of a representative user. Given the definition of the mean field $m(x, t)$ in (5.11), the running cost function (5.7) becomes

$$r(x, m, u) = c_1 u^2 + c_2 m + c_3 \lambda \, \|x - x_d\| + c_4 (1 - \lambda) \, \|x - \bar{x}_0\| , \qquad (5.14)$$

where $\bar{x}_0$ denotes the average initial opinion and is different from x_{i0} in (5.7) which corresponds to the initial opinion of player i.

The first term of the running cost function refers to the cost associated with control effort. In the literature, control effort usually has quadratic form to denote diminishing return. That is, the cost incurred by an additional unit of control effort increases with increasing control effort. The quadratic form is also desirable since it yields tractable solution due to a convex Hamiltonian $H(x, m, p)$. The second term refers to the effect of the mean field of the population to the cost incurred by a user. For small c_2, the cost of a player decreases with the density. This means that a popular opinion is more desirable. For large c_2, the cost of a user increases with the density. This means that an unpopular opinion is more desirable. The third term refers to the difference of the user's opinion x to the desired state x_d. It is penalized so that the user will try to find solution that minimizes its opinion distance with x_d. The fourth term refers to the stubbornness of the user. That is, it refers to the degree at which the user prefers its initial opinion x_0. The constant λ refers to the compromise of a user between its initial opinion x_0 and its desired opinion x_d. The desired opinion x_d is set according to the dataset. It represents the target or ideal opinion of the population in the dataset. The subscript i has been dropped to imply that the users are indistinguishable from each other and that they use the same cost function.

Meanwhile, the terminal cost function g of a social network user is neglected since the user is not required to have the desired state at time $t = T$. Consequently, the average total cost of a user is given by

$$J(u, m) = \int_0^T \int_{\mathscr{X}} (c_1 u^2 + c_2 m + c_3 \lambda \, \|x - x_d\| + c_4(1 - \lambda) \, \|x - \bar{x}_0\|) m(x, t) \, dx \, dt.$$

$$(5.15)$$

To be able to derive the opinion dynamics in terms of the mean field $m(x, t)$ and the population mean $\bar{x}$, the strong law of large numbers (SLLN) [6], an important concept in probability theory, is referenced.

Theorem 5.2 *Let* $x_1, x_2, \ldots, x_n$ *be a sequence of independent and identically distributed random variables, each having the same finite mean* μ. *Then,*

$$\mathbb{P}\left(\lim_{N \to \infty} \frac{1}{N}(x_1 + x_2 + \cdots + x_N) = \mu \right) = 1.$$

In words, the partial averages $\frac{1}{N}(x_1 + x_2 + \cdots + x_N)$ *converges almost surely to* μ.

In (5.3), let $\bar{x}_n(k) = \left(\sum_{j=1}^N \alpha_{ij} x_j(k) \right) / \sum_{j=1}^N \alpha_{ij}$ refer to the neighborhood mean of user i. Then (5.3) can be written as

$$x_i(k + 1) = \lambda_i \bar{x}_n(k) + (1 - \lambda_i)x_{i0} + \beta_i u_i(x_i, k).$$

It can be deduced that, at any time $k + 1$, the opinion of user i is independent of the opinion of other users at time $k + 1$ and dependent only on the opinion of other users at time k. Hence, the SLLN can be used to extend the neighborhood mean opinion $\bar{x}_n$ to population mean opinion $\bar{x}$.

Based on Theorem 5.2, the population mean $\bar{x}$ and neighborhood mean $\bar{x}_n$ can be connected as $\bar{x}_n = \alpha \bar{x}$, where α is a constant. As $N \to \infty$, $\bar{x}_n \to \bar{x}$, and thus $\alpha \to 1$. Consequently, the opinion dynamics in (5.4) can be written in terms of the population mean, $\bar{x}$

$$dx(t) = \bar{a}\left(\lambda \int_{\mathscr{X}} xm\,dx + (1-\lambda)\int_{\mathscr{X}} xm_0\,dx\right)dt + (ax+bu)\,dt + \sigma\,dw(t),$$

$$= \big(\bar{a}(\lambda\bar{x} + (1-\lambda)\bar{x}_0) + ax + bu\big)\,dt + \sigma\,dw(t),$$

$$= f(x,m,u)\,dt + \sigma\,dw(t),$$

$$(5.16)$$

where $\bar{x}$ is the average population opinion, $\bar{x}_n = \sum_{j=1}^{N}\alpha_{ij}x_j / \sum_{j=1}^{N}\alpha_{ij}$ is the average neighborhood opinion, and $\bar{a}$, a, and b are constants. The third term in (5.16) is added to capture the opinion change due to uncertainty or random phenomenon with σ as the diffusion constant and $w(t)$ as a standard Wiener process. A discrete-time equation can be converted into a continuous-time equation using $\bar{x}(t) = \bar{x}(hk) = \bar{x}(k)$ with $h \to \epsilon$, where $\epsilon > 0$ is a very small number [7].

Figure 5.1b shows an illustration of a single-population MFG social network, where the aggregate effect of other users to a representative user is taken into account rather than the individual effect of each other user. The social network problem solved through an MFG approach is formally stated as follows.

Problem 5.1 Consider a social network with a large number of indistinguishable users. In this network, every user behaves rationally by minimizing the total cost $J(u,m)$ associated to its preferences given the opinion dynamics $dx(t)$ it experiences within the network. Let $x \in \mathbb{R}$ be the scalar opinion of a representative user. Denote $v(x,t) = \min_{u\in\mathscr{U}} J(u,m)$ and $u(x,t)$ as the value and control functions of the user, respectively. Let $m(x,t)$ correspond to the distribution of users according to opinion x at time t. Then, any user views the network as a mean field game, where the dynamics of $v(x,t)$ and $m(x,t)$ with time are represented by the pair of HJB and FPK equations,

$$-\frac{\partial v}{\partial t}(x,t) - H(x,m,p) = \frac{\sigma^2}{2}\frac{\partial^2 v}{\partial x^2}(x,t),$$

$$\frac{\partial m}{\partial t}(x,t) + \mathrm{div}\left(\frac{\partial H}{\partial p}(x,m,p)m(x,t)\right) = \frac{\sigma^2}{2}\frac{\partial^2 m}{\partial x^2}(x,t),$$

with boundary conditions $v(x,T) = g(x,m_T)$ and $m(x,0) = m_0(x)$, where $m_T = m(x,T)$. Meanwhile, the optimal control u^* is the solution to the equation

$$f(x,m,u^*) = \frac{\partial H}{\partial p}(x,m,p),$$

where the Hamiltonian is defined as $H(x,m,p) = \min_{u\in\mathscr{U}} r(x,u,m) + f(x,u,m)p(t)$ and $p(t) = \frac{\partial v}{\partial x}(x,t)$. The running cost $r(x,u,m)$ and drift $f(x,u,m)$ are given by

$$r(x,u,m) = c_1 u^2 + c_2 m + c_3\lambda\,\|x - x_d\| + c_4(1-\lambda)\,\|x - \bar{x}_0\|,$$

$$f(x,u,m) = \bar{a}(\lambda\bar{x} + (1-\lambda)\bar{x}_0) + ax + bu,$$

where λ denotes the average susceptibility to neighbor influence.

In a single population, the MFG focuses on the Nash equilibrium (i.e., a solution in which no player gains anything by changing its own strategy) among the players in the group. The same point of view can be extended to scenarios with several populations. In the following section, an extension of the MFG formulation to scenarios involving several populations is discussed. Then, the methodology is applied to social networks where users can be grouped into several populations.

5.3 Mean Field Game with Several Populations

In this section, the multiple-population MFG (MPMFG) framework presented in [8] is introduced. This MFG framework will be used in the next section to model the belief and opinion dynamics in multiple-population social networks.

5.3.1 Background and Motivation

Consider a non-cooperative game among a large number of players that can be divided into P populations. Let the state $x \in \mathcal{X}$. Denote mean field vector $m = (m_i)_{i=1,...,P}^\top$ and control vector $u = (u_i)_{i=1,...,P}^\top$. Assume that each population considers that the distribution of all populations are fixed or predictable and tries to minimize its own individual cost. The goal of a player in population i is to minimize the population cost $J_i(u_i, m)$ subject to the population state dynamics $dx(t)$.

In an MPMFG problem with P populations, population i faces the problem of finding a pair (u_i^*, m_i^*) that solves

$$\min_{u_i \in \mathcal{U}_i} J_i(u_i, m) = \mathbb{E}\left[\int_0^T r_i(x, u_i, m)\, dt + g_i(x_T, m_T)\right], \qquad (5.17)$$

subject to the mean field dynamics

$$\frac{\partial m_i}{\partial t}(x, t) + \mathrm{div}\left[f_i(x, u_i, m)m_i(x, t)\right] = \frac{\sigma_i^2}{2}\Delta_x m_i(x, t),$$

$$\frac{\partial m_l}{\partial t}(x, t) + \mathrm{div}\left[f_l(x, u_l^*, m)m_l(x, t)\right] = \frac{\sigma_l^2}{2}\Delta_x m_l(x, t), \qquad (5.18)$$

for $l \neq i$, with boundary conditions $m_i(x, 0) = m_{i,0}(x)$ and $m(x, T) = m_T$, where $x_T = x(T)$, for all P populations.

The pair (u_i^*, m_i^*) that solves the MPMFG problem above is also a solution of the HJB and FPK equations for population i. Hence, the mean field m_i of the MPMFG problem must coincide with the mean field m_i^* solution of the FPK equation for population i.

The following theorem is the formal expression of this MFGs with several populations. It is an extension of Theorem 2.3 when there are multiple interacting populations, each with large number of indistinguishable players. For the detailed proof, please refer to [8] and the references therein.

Theorem 5.3 *Consider a non-cooperative game among large a number of indistinguishable players. If population i faces the optimization problem*

$$\min_{u_i \in \mathscr{U}_i} J_i(u_i, m) = \mathbb{E}\left[\int_0^T r_i(x, u_i, m)\, dt + g_i(x_T, m_T) \right], \tag{5.19}$$

subject to

$$\frac{\partial m_i}{\partial t}(x, t) + div\big[f_i(x, u_i, m)m_i(x, t) \big] = \frac{\sigma_i^2}{2} \Delta_x m_i(x, t),$$
$$\frac{\partial m_l}{\partial t}(x, t) + div\big[f_l(x, u_l^*, m)m_l(x, t) \big] = \frac{\sigma_l^2}{2} \Delta_x m_l(x, t), \tag{5.20}$$

for $l \neq i$, then the equivalent non-cooperative mean field game consists of the HJB and FPK equations,

$$-\frac{\partial v_i}{\partial t}(x, t) - H_i(x, m, p_i) = \frac{\sigma_i^2}{2} \Delta_x v_i(x, t),$$
$$\frac{\partial m_i}{\partial t}(x, t) + div\left(\frac{\partial H_i}{\partial p_i}(x, m, p_i)m_i(x, t) \right) = \frac{\sigma_i^2}{2} \Delta_x m_i(x, t), \tag{5.21}$$

with boundary conditions $v_i(x, T) = g_i(x, m_T)$ and $m_i(x, 0) = m_{i,0}(x)$, where $m_T = m(x, T)$ and $x_T = x(T)$, for all P populations. The Hamiltonian $H_i(x, m, p_i)$ is defined as

$$H_i(x, m, p_i) = \min_{u_i \in \mathscr{U}_i} r_i(x, u_i, m) + f_i(x, u_i, m) \cdot p_i(t), \tag{5.22}$$

with $p_i(t) = \frac{\partial v_i}{\partial x}(x, t)$.
Meanwhile, the optimal control u_i^* is the solution to the equation

$$f_i(x, u_i^*, m) = \frac{\partial H_i}{\partial p_i}(x, m, p_i),$$

such that the mean field m_i when $u_i = u_i^*$ in (5.19)–(5.20) coincides with the solution m_i^* of the FPK equation in (5.21).

5.3.2   A Mean Field Game Problem for Multiple-Population Social Networks

In this subsection, the MPMFG model in the previous section is applied to a social network consisting of multiple populations where the users can be divided into several populations and the users in the same population or group share common characteristics.

Let $x \in \mathbb{R}$ be the scalar opinion of a representative user in a population. The running cost $r_i(x, u_i, m)$ of a social network user in population i is proportional to the influence effort $u_i(x, t)$ it exerts to minimize its own cost. In addition, a user is rewarded (or penalized) if it influences users from any population. In other words, the cost decreases (or increases) if the mean fields of the populations $m_l, l = 1, \ldots, P$ are higher (or lower). If x_{id} denotes the ideal or desired opinion of population i, then each user must minimize the distance of its opinion to the ideal opinion, $\|x - x_{id}\|$. In addition, when the social network users have a degree of stubbornness, the distance to the average initial opinion also affects the cost, $\|x - \bar{x}_{i0}\|$. Therefore, the running cost function r_i in (5.14) can be written mathematically as

$$r_i(x, u_i, m) = c_1 u_i^2 + \sum_{l=1}^{P} c_{2il} m_l + c_3 \lambda_i \|x - x_{id}\| + c_4(1 - \lambda_i) \|x - \bar{x}_{i0}\|.$$
(5.23)

Meanwhile, the terminal cost function g_i of a social network user in population i is neglected since the user is not required to have the desired state of population i or the users are not required to have a desired mean field at time $t = T$. Consequently, the average total cost of a user in population i is given by

$$J_i(u_i, m) = \int_0^T \int_{\mathscr{X}} \left(c_1 u_i^2 + \sum_{l=1}^{P} c_{2il} m_l + c_3 \lambda_i \|x - x_{id}\| \right.$$

$$\left. + c_4(1 - \lambda_i) \|x - \bar{x}_{i0}\| \right) m_i(x, t) \, dx \, dt.$$
(5.24)

Since the users can be divided into P distinct populations, the opinion dynamics equation in (5.16) for population i becomes

$$dx(t) = \left(\lambda_i \sum_{l=1}^{P} \eta_l \bar{a}_{il} \bar{x}_l + (1 - \lambda_i) \bar{a}_{ii} \bar{x}_{i0} + a_i x + b_i u_i \right) dt + \sigma_i \, dw_i(t),$$

$$= f_i(x, u_i, m) \, dt + \sigma_i \, dw_i(t),$$
(5.25)

where η_l refers to the ratio of the size of population l and the total population, $\bar{x}_l = \bar{x}_{nl}/\alpha_{il}$ is the average opinion at population l, α_{il} is the average adjacency to population l in population i, $\bar{a}_{il} = \bar{a}/\alpha_{il}$, b_i is a control constant, σ_i is the diffusion constant, and $w_i(t)$ is a standard Wiener process. In addition, λ_i is the average susceptibility of a user in population i.

The following problem states the multiple-population opinion evolution on social network modeled using MFGs.

Problem 5.2 Consider a social network of P distinct populations, each with a large number of indistinguishable users sharing a common characteristic. In this network, every user in population i behaves rationally by minimizing the total cost $J_i(u_i, m)$ associated to its preferences given the opinion dynamics $dx(t)$ it experiences within the network. Let $x \in \mathbb{R}$ be the scalar opinion of a representative user in a population. Denote $v_i(x, t) = \min_{u_i \in \mathcal{U}_i} J_i(u_i, m)$ and $u_i(x, t)$ as the value and control functions of a user in population i, respectively. Let $m_i(x, t)$ correspond to the distribution of user at population i according to opinion x at time t. Then, any user at population i views the network as an MPMFG where the evolutions of $v_i(x, t)$ and $m_i(x, t)$ with time are represented by the pair of HJB and FPK equations,

$$-\frac{\partial v_i}{\partial t}(x, t) - H_i(x, m, p_i) = \frac{\sigma_i^2}{2} \frac{\partial^2 v_i}{\partial x^2}(x, t),$$

$$\frac{\partial m_i}{\partial t}(x, t) + \text{div}\left(\frac{\partial H_i}{\partial p_i}(x, m, p_i) m_i(x, t)\right) = \frac{\sigma_i^2}{2} \frac{\partial^2 m_i}{\partial x^2} m_i(x, t),$$

with boundary conditions $v_i(x, T) = g_i(x, m_T)$ and $m_i(x, 0) = m_{i,0}(x)$, where $m_T = m(x, T)$. Meanwhile, the optimal control u_i^* of a user in population i is the solution to the equation

$$f_i(x, u_i^*, m) = \frac{\partial H_i}{\partial p_i}(x, m, p_i),$$

where the Hamiltonian is defined as $H_i(x, m, p_i) = \min_{u_i \in \mathcal{U}_i} r_i(x, u_i, m) + f_i(x, u_i, m) p_i(t)$ and $p_i(t) = \frac{\partial v_i}{\partial x}(x, t)$. The running cost $r_i(x, u_i, m)$ and drift $f_i(x, u_i, m)$ of population i are given by

$$r_i(x, u_i, m) = c_1 u_i^2 + \sum_{l=1}^{P} c_{2il} m_l + c_3 \lambda_i \|x - x_{id}\| + c_4(1 - \lambda_i) \|x - \bar{x}_{i0}\|,$$

$$f_i(x, u_i, m) = \lambda_i \sum_{l=1}^{P} \eta_l \bar{a}_{il} \bar{x}_l + (1 - \lambda_i) \bar{a}_{ii} \bar{x}_{i0} + a_i x + b_i u_i,$$

where λ_i refers to the average susceptibility to neighbor influence in population i.

In the next section, an analytical method to solve the pair of partial differential equations in (5.21) is presented. Then, this method is applied to the proposed social network problem.

5.4 Adjoint Method for Mean Field Games with Several Populations

The adjoint method is a method of solving optimization problems with partial differential equation (PDE) constraints [9]. For single-population MFG, an adjoint method was presented in [10] to solve the MFG problem with PDE constraint. In this section, this methodology is extended to MPMFGs and the HJB-FPK pair (5.21) are solved for population i. Afterwards, the application of the methodology to the social network belief and opinion evolution problems is presented.

5.4.1 Background

The method starts with introducing an adjoint variable, $v_i(x, t)$, which also corresponds to the value function

$$v_i(x, t) = \min_{u_i \in \mathcal{U}_i} J_i(u_i, m). \tag{5.26}$$

Using this adjoint variable, the FPK equation constraints are appended to the original cost function $J_i(u_i, m)$.

Consider the optimal control problem

$$\min_{u_i \in \mathcal{U}_i} J_i(u_i, m) = \mathbb{E}\left[\int_0^T r_i(x, u_i, m) \, dt + g_i(x_T, m_T) \right], \tag{5.27}$$

subject to

$$\frac{\partial m_l}{\partial t}(x, t) + \text{div}\left[f_l(x, u_l, m) m_l(x, t) \right] = \frac{\sigma_l^2}{2} \Delta_x m_l(x, t), \tag{5.28}$$

for $l = 1, \ldots, P$. The resulting MPMFG problem optimizes the extended cost function $\mathcal{J}_i$

$$\mathcal{J}_i(u_i, m_i, v_i) = \int_0^T \int_{\mathcal{X}} r_i(x, u_i, m) m_i(x, t) \, dx \, dt + \int_{\mathcal{X}} g_i(x_T, m_T) m_i(x, T) \, dx$$

$$+ \sum_{l=1}^P \int_0^T \int_{\mathcal{X}} v_l(x, t) \left(-\frac{\partial m_l}{\partial t}(x, t) - \text{div}\left[f_l(x, u_l, m) m_l(x, t) \right] \right.$$

$$\left. + \frac{\sigma_l^2}{2} \Delta_x m_l(x, t) \right) dx \, dt$$

$$\tag{5.29}$$

There exists a pair (m_i, u_i) that minimizes the extended cost function in (5.29) if there is a v_i such that (m_i, u_i, v_i) is a stationary solution. Thus, the following optimality conditions are necessary:

$$\frac{\partial \mathscr{J}_i}{\partial u_i} = 0, \quad \frac{\partial \mathscr{J}_i}{\partial m_i} = 0, \quad \text{and} \quad \frac{\partial \mathscr{J}_i}{\partial v_i} = 0. \tag{5.30}$$

Solving $\frac{\partial \mathscr{J}_i}{\partial u_i} = 0$ yields

$$\frac{\partial r_i}{\partial u_i}(x, u_i, m) + p_i(t)\frac{\partial f_i}{\partial u_i}(x, u_i, m) + \frac{\partial}{\partial u_i}\left(\frac{\sigma_i^2}{2}\Delta_x v_i\right) = 0, \tag{5.31}$$

where $p_i(t) = \nabla_x v_i(x, t)$.

Meanwhile, $\frac{\partial \mathscr{J}_i}{\partial m_i} = 0$ results to

$$-\frac{\partial v_i}{\partial t}(x, t) - H_i(x, m, p_i) = \frac{\sigma_i^2}{2}\Delta_x v_i(x, t)$$

$$+ m_i \frac{\partial r_i}{\partial m_i}(x, u_i, m) + \sum_{l=1}^{P}\left(m_l\frac{\partial f_l}{\partial m_i}(x, u_l, m)\nabla_x v_l(x, t)\right). \tag{5.32}$$

Finally, $\frac{\partial \mathscr{J}_i}{\partial v_i} = 0$ yields

$$\frac{\partial m_i}{\partial t}(x, t) + \text{div}\big[f_i(x, u_i, m)m_i(x, t)\big] = \frac{\sigma_i^2}{2}\Delta_x m_i(x, t). \tag{5.33}$$

In summary, (5.31)–(5.33) are solved iteratively to find the stationary solution (m_i, u_i, v_i).

5.4.2 Adjoint Method for the Social Network Problems

For the social network problem, applying the optimality conditions in (5.30) to the extended cost function $\mathscr{J}_i(m_i, u_i, v_i)$ leads to the following equations:

$$\frac{\partial \mathscr{J}_i}{\partial u_i} = m_i\frac{\partial r_i}{\partial u_i}(x, u_i, m) + bm_i\frac{\partial v_i}{\partial x}(x, t), \tag{5.34}$$

$$\frac{\partial \mathscr{J}_i}{\partial m_i} = \frac{\partial v_i}{\partial t}(x, t) + H_i(x, m, p_i) + \frac{\sigma_i^2}{2}\frac{\partial^2 v_i}{\partial x^2}(x, t) + c_{2i}m_i$$

$$+ \frac{1}{2}\big(\lambda_i\eta_i\bar{a}_{ii}x^2 + (1 - \lambda_i)\bar{a}_{ii}x^2\big)\sum_{l=1}^{P}m_l\frac{\partial v_l}{\partial x}(x, t), \tag{5.35}$$

$$\frac{\partial \mathscr{J}_i}{\partial v_i} = \frac{\partial m_i}{\partial t}(x,t) + \text{div}\big[f_i(x,u_i,m)m_i(x,t)\big] - \frac{\sigma_i^2}{2}\frac{\partial^2 m_i}{\partial x^2}(x,t). \qquad (5.36)$$

In the next section, a numerical method to solve the optimality conditions in (5.31)–(5.33) is presented.

5.5 Numerical Method for Mean Field Games with Several Populations

To calculate the stationary solution (m_i, u_i, v_i) corresponding to the optimality conditions in (5.31)–(5.33), a well-known numerical method called finite difference method is implemented [11]. In this method, the PDEs are converted into their equivalent discrete-time difference equations. In addition to finite difference method, a numerical scheme called the Lax-Friedrichs scheme is discussed. Afterwards, these numerical techniques are applied to solve the MFG problems presented in the previous sections.

5.5.1 Background

Consider a bounded region $[0, X_{\max}] \times [0, T_{\max}]$ over which independent variables x and t of the PDEs are defined. In order to implement a numerical method to solve the PDEs, the region is converted into a finite grid of points. Given positive integers L and M, a space step is defined as $\Delta x = \frac{X_{\max}}{L}$, and a time step is defined as $\Delta t = \frac{T_{\max}}{M}$. Hence, the grid of points are defined by

$$x_j = j\Delta x, j = 0, ..., L, \qquad (5.37)$$

and
$$t_k = k\Delta t, k = 0, ..., M. \qquad (5.38)$$

Furthermore, for any function ρ defined over the space-time region, $\rho_j^k = \rho(x_j, t_k)$.

The finite difference (FD) method is a numerical method to approximate and solve PDEs. This method uses finite differences to approximate derivatives. For the first-order derivatives in x, the forward, backward, and central differences are defined as

$$\frac{\partial \rho}{\partial x} = \frac{\rho_{j+1}^k - \rho_j^k}{\Delta x}, \qquad (5.39)$$

$$\frac{\partial \rho}{\partial x} = \frac{\rho_j^k - \rho_{j-1}^k}{\Delta x}, \qquad (5.40)$$

and
$$\frac{\partial \rho}{\partial x} = \frac{\rho_{j+1}^k - \rho_{j-1}^k}{2\Delta x},$$
(5.41)

where $\Delta x = x_{j+1} - x_j = x_j - x_{j-1}$. For the second-order derivatives in x, the symmetric difference approximation is defined as

$$\frac{\partial^2 \rho}{\partial x^2} = \frac{\rho_{j+1}^k - 2\rho_j^k + \rho_{j-1}^k}{(\Delta x)^2}.$$
(5.42)

These definitions apply to derivatives in t as well.

Applying the FD method to the mean field $m_i(x, t)$ results to

$$\frac{m_{i,j}^{k+1} - m_{i,j}^k}{\Delta t} + \frac{\rho_i(m_{i,j}^k, m_{i,j+1}^k) - \rho_i(m_{i,j-1}^k, m_{i,j}^k)}{\Delta x} = \varepsilon \frac{m_{i,j+1}^k - 2m_{i,j}^k + m_{i,j-1}^k}{(\Delta x)^2},$$
(5.43)

where $\mathrm{div}\big[f_i(x, m, u_i)m_i(x, t)\big]$ has been expressed in terms of a consistent numerical flow $\rho_{i,j}^k$.

To preserve the sign of $m_{i,j}^k$ in (5.43), a monotone numerical scheme such as the Lax-Friedrichs scheme should be utilized. Given the Lax-Friedrichs flux

$$\rho(m, n) = \frac{1}{2}(\phi(m) + \phi(n)) + \frac{\Delta t}{2\Delta x}(m - n),$$
(5.44)

the Lax-Friedrichs scheme for (5.43) is

$$m_j^{k+1} = \frac{1}{2}(m_{j+1}^k + m_{j-1}^k) - \frac{\Delta t}{2\Delta x}(\phi_{j+1}^k - \phi_{j-1}^k) + \varepsilon \frac{\Delta t}{(\Delta x)^2}(m_{j+1}^k - 2m_j^k + m_{j-1}^k),$$
(5.45)

where $\phi_j^k = f_j^k m_j^k$.

Consequently, the discrete form of $\mathscr{J}_i(m_i, u_i, v_i)$ is

$$\mathscr{J}_i(m_{i,j}^k, u_{i,j}^k, v_{i,j}^k) = \Delta x \Delta t \sum_{j,k}^{L,M} r_{i,j}^k m_{i,j}^k + \Delta x \sum_j^L g_j^N$$

$$+ \Delta x \Delta t \sum_{l=1}^P \sum_{j,k}^{L,M} -v_{l,j}^k \frac{m_{l,j}^{k+1} - 0.5(m_{l,j+1}^k + m_{l,j-1}^k)}{\Delta t}$$

$$- v_{l,j}^k \frac{\phi_{l,j+1}^k - \phi_{l,j-1}^k}{2\Delta x} + \varepsilon v_{l,j}^k \frac{m_{l,j+1}^k - 2m_{l,j}^k + m_{l,j-1}^k}{(\Delta x)^2}.$$
(5.46)

Algorithm 1: Solving the mean field game problem for several populations

---

1: **Initialize:** $m_{i,j}^0, v_{i,j}^M, u_{i,j}^0,$
2: **while** it $\leq I$ or error $\leq \epsilon$ **do**
3:     **for** $i = 1, \ldots, P, j = 0, \ldots, L, k = 0, \ldots, M-1$ **do**
4:         Solve the mean field $m_{i,j}^{k+1}$ using (5.47).
5:     **end for**
6:     **for** $i = 1, \ldots, P, j = 0, \ldots, L, k = M, \ldots, 2$ **do**
7:         Solve the value $v_{i,j}^{k-1}$ using (5.48).
8:     **end for**
9:     **for** $i = 1, \ldots, P, j = 0, \ldots, L, k = 0, \ldots, M$ **do**
10:        Update $u_{i,j}^k$ using (5.49).
11:     **end for**
12: **end while**

5.5.2 Numerical Method for the Social Network Problems

After applying the optimality conditions in (5.30) to (5.46), the update rules for the mean field $m_{i,j}^{k+1}$, value $v_{i,j}^{k-1}$ and control $u_{i,j}^k$ can be solved.

The mean field $m_{i,j}^{k+1}$ for the social network problems is in (5.47). The update rule for the value function is in (5.48). Lastly, the following update rule for the control function $u_{i,j}^k$ is in (5.49), where w is a constant.

$$m_{i,j}^{k+1} = 0.5(m_{i,j+1}^k + m_{i,j-1}^k) + \Delta t\left(-\frac{\phi_{i,j+1}^k - \phi_{i,j-1}^k}{2\Delta x} + \varepsilon \frac{m_{i,j+1}^k - 2m_{i,j}^k + m_{i,j-1}^k}{(\Delta x)^2} \right)$$

$$(5.47)$$

$$v_{i,j}^{k-1} = 0.5(v_{i,j-1}^k + v_{i,j+1}^k) + \Delta t\left(r_{i,j}^k + m_{i,j}^k \frac{\partial r_{i,j}^k}{m_{i,j}^k} - f_{i,j}^k \frac{v_{i,j-1}^k - v_{i,j+1}^k}{2\Delta x} \right.$$

$$\left. - \frac{\partial f_i}{m_{i,j}^k} \sum_{l=1}^{P} m_{l,j}^k \frac{v_{i,j-1}^k - v_{i,j+1}^k}{2\Delta x} + \varepsilon \frac{v_{i,j-1}^k - 2v_{i,j}^k + v_{i,j+1}^k}{(\Delta x)^2} \right) \qquad (5.48)$$

$$u_{i,j}^k \leftarrow \frac{w}{1+w} u_{i,j}^k - \frac{1}{1+w} \frac{\partial r_i}{\partial u_{i,j}^k} \qquad (5.49)$$

Algorithm 1 provides the procedure of solving (5.47)–(5.49) for the multiple-population social network belief and opinion evolution. In the algorithm, I is the maximum number of iterations and w is the control update parameter.

5.6   Simulation Results and Discussion

5.6.1   Theoretical Results

The following simulations demonstrate the theoretical aspects of the evolution of opinion in a social network consisting of single and multiple groups of users. The mean field $m_i(x, t)$ refers to the distribution of opinion x at population i at time t. To calculate the solution (m_i, v_i, u_i) of the MFG-based belief and opinion evolution models, Algorithm 1 is implemented using MATLAB.

Consider a social network consisting of users that can be divided into two separate populations, $i = 1, 2$. Let the state or opinion of a user be $x \in [0, 1]$ and time be $t \in [0, 1]$. Let the initial mean fields $m_i(x, 0)$ be $\mathcal{N}(0.25, 0.1)$ and $\mathcal{N}(0.75, 0.1)$ for populations 1 and 2, respectively. Also, assume a 50%–50% ratio between the two populations (i.e., $\eta_1 = \eta_2 = 0.5$). The opinion dynamics equation parameters are set at $\bar{a}_i = 0.001$, $a_i = -0.001$, and $\sigma_i = 0.01$. Meanwhile, the cost function parameters are $c_1 = 0.5$, $c_2 = 1$, $c_3 = 2$, $c_4 = 1$, $x_{01} = 0.25$, $x_{02} = 0.75$. Finally, $w = 1000$, $I = 2000$ and $\epsilon = 1 \times 10^{-6}$.

Figure 5.2 shows the mean field $m_i(x, t)$ at $t = 0, 0.5, 1$ for populations $i = 1, 2$ in blue and red, respectively: (a) independent and formulated as two single-population MFG; and (b) dependent and formulated as a two-population MFG. Based on the figure, the presence of another population affects the distribution of opinion of the other population. Specifically, the populations hold on to the their initial distributions with much ease in the presence of a competing population.

The effect of social network user stubbornness is presented in Fig. 5.3. When users are more susceptible to neighbor opinions, or have low stubbornness, more users are likely to achieve the population target or ideal opinion, as shown in Fig. 5.3a with $\lambda_i = 0.9$. However, as users become less susceptible to neighbor opinions, or have high stubbornness, more users are likely to stay with the initial population opinion, as shown in Fig. 5.3b with $\lambda_i = 0.1$.

The effect of other parameters to the mean field $m_i(x, t)$ of population i are shown in Fig. 5.4. In Fig. 5.4a, population $i = 1$ with a higher percentage (i.e., 90% of total population), achieves a larger variance than population $i = 2$ with a lower percentage (i.e., 10% of total population). Meanwhile, in Fig. 5.4b, the distance between the means μ_i introduces variations in the mean field $m_i(x, t)$, as shown by the asymmetrical plots of $m_i(x, t)$.

5.6.2   Experimental Results with Real Dataset

The following simulations test the effectiveness of the proposed MFG-based algorithm in modeling the opinion evolution of a social network with multiple populations. The experiments use the social evolution dataset published in [12].

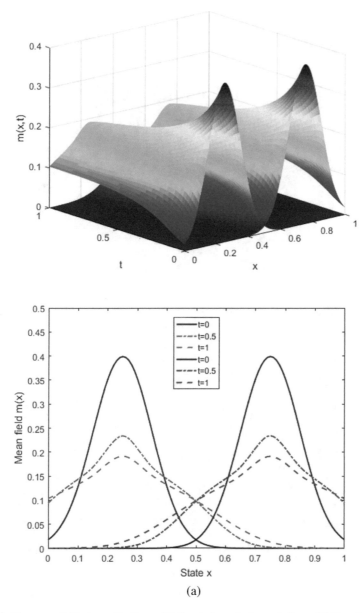

Fig. 5.2 The mean field at specific times. (**a**) Two independent populations. (**b**) Two dependent populations

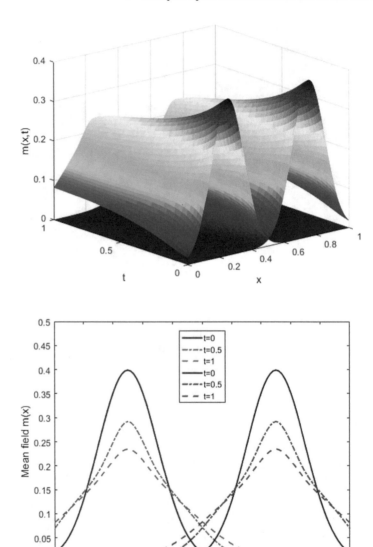

Fig. 5.2 (continued)

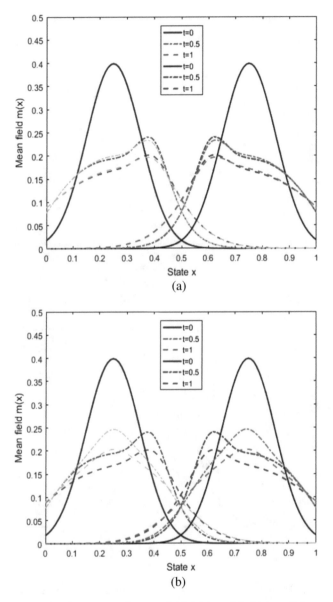

Fig. 5.3 The effect of stubbornness to the mean field. (**a**) $\lambda_i = 0.9$. (**b**) $\lambda_i = 0.1$

5.6.2.1 Description of the Dataset

The social evolution dataset consists of the political parties and opinions of 80 subjects as well as the relationship among them. The subjects were college students living in the same housing facility at MIT. One of the attributes of the subjects

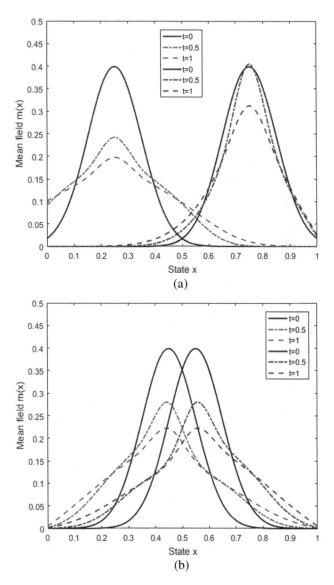

Fig. 5.4 The effect of population fraction and mean state to the mean field. (**a**) $\eta_1 = 0.9$, $\eta_2 = 0.1$. (**b**) $\mu_1 = 0.45$, $\mu_2 = 0.55$

is their political party or affiliation (i.e., *Democratic Party* or *Republican Party*). The students were surveyed three times, March, April, and June 2009, about their opinion (i.e., approval or satisfaction level) of U.S. President Obama as the head of the state and the government. In each survey, they were asked whether they *"strongly disapprove," "disapprove," "don't know," "approve,"* or *"strongly approve,"* of

President Obama. Moreover, they were asked about the extent of their relationship with other students involved in the survey. A relationship between two students exists if they are close friends, have had political discussions, and are involved together in social activities as well as social media networks such as Facebook, Twitter, blog and live journal posts.

5.6.2.2 Processing of the Dataset

To be able to utilize the dataset for the proposed MFG-based method, the opinions and relationship links are converted into their numerical equivalent. For the state or opinion x of a student, "*strongly disapprove*" $= 0.1$, "*disapprove*" $= 0.3$, "*don't know*" $= 0.5$, "*approve*" $= 0.7$, and "*strongly approve*" $= 0.9$. Meanwhile, the control or number of interactions u of a student refers to the total number of relationships it has had during the month. Then, the number of relationships are normalized between $[u_{min}, u_{max}]$ and centered at $u = 0$.

The state dynamics and cost function parameters are calculated and estimated from the dataset. The expectation of the state dynamics for political party i in (5.16) can be rewritten as

$$\mathbb{E}\left[\frac{dx}{dt}\right] = \mathbb{E}[\dot{x}] = \mathbb{E}\left[\bar{a}_i(\lambda_i\bar{x} + (1 - \lambda_i)\bar{x}_{i0}) + a_i x + b_i u_i\right]$$

$$= \mathbb{E}\left[\omega_s \dot{x} + \omega_c \dot{x}\right], \tag{5.50}$$

where $\bar{x}$ is the average opinion of all the students, $\bar{x}_{i0}$ is the average opinion of students at $t = 0$, ω_s is the portion of $\dot{x}$ due to state x and ω_c is the portion of $\dot{x}$ due to control u_i. Note that, from the definition of a standard Wiener process, $\mathbb{E}[dw_i(t)] = 0$. Thus, the parameters of the state dynamics in (5.16) can be calculated from the dataset as

$$\bar{a}_i = \frac{\omega_s \mathbb{E}[\dot{x}]}{\mathbb{E}[\lambda_i\bar{x} + (1 - \lambda_i)\bar{x}_{i0} - x]} \quad \text{and} \quad b_i = \frac{\omega_c \mathbb{E}[\dot{x}]}{\mathbb{E}[u_i]}, \tag{5.51}$$

where the expectation is taken with respect to all the students in the survey. Since the dataset contains only the political opinions of the students, the values of the cost function parameters in (5.23) are estimated until the theoretical mean field $m_i(x, 1)$ fits the calculated mean field $M_i(x, t)$ from the dataset.

5.6.2.3 Procedure of the Experiment

In the following simulations, $i = 1$ for the Democratic Party and $i = 2$ for the Republican Party, while $t = 0$ for March 2009, $t = 1$ for April 2009, and $t = 2$ for June 2009. Given the political opinions for March, April, and June 2009, the distribution $M_i(x, t)$ of opinions of political party i is calculated for each month.

The goal of the simulations is to use the MFG-based algorithm to model and predict the evolution of distribution of opinions of the students regarding President Obama. Hence, given $M_i(x, 0)$, the mean field $m_i(x, 1)$ is predicted using the proposed MFG-based model and compared with the calculated $M_i(x, 1)$. Similarly, given $M_i(x, 1)$, the mean field $m_i(x, 2)$ is predicted and compared with the calculated $M_i(x, 2)$.

5.6.2.4   Results and Analysis

Figure 5.5 shows the results of estimating the opinion distribution for the Democratic Party (DP) and Republican Party (RP) from the dataset. The performance is evaluated by calculating the average absolute error between the estimated mean field $m_i(x, t)$ from the proposed MFG-based algorithm and the calculated mean field $M_i(x, t)$ from the dataset. In Fig. 5.5a, the data for March 2009 is used to estimate the distribution of opinions for April 2009. The accuracy of the estimation is 97.25%. Meanwhile, in Fig. 5.5b, the data for April 2009 is used to estimate the distribution of opinions for June 2009. The accuracy of the estimation is 96.93%.

Finally, Fig. 5.6 presents the prediction of opinion distribution for May and August 2009. The prediction for May 2009 uses the data collected during April 2009, as shown in Fig. 5.6a, while the prediction for August 2009 uses the data collected during June 2009, as shown in Fig. 5.6b. It can be concluded from these results that the distribution of opinions are affected by the social dynamics (i.e., opinion change and relationships) of the students.

5.6.3   Performance Analysis

In this subsection, the validity and effectiveness of the proposed MFG opinion model are studied by comparing the MFG model with the classical opinion dynamics models. The performance metric used to measure the performance of the proposed MFG model is the relative entropy. It is calculates the difference between the measured probability distribution P and a reference probability distribution R. Mathematically, the relative entropy is given by the equation

$$H(P||R) = \int_{\mathscr{X}} p(x) \ln \left(\frac{p(x)}{r(x)} \right) dx, \qquad (5.52)$$

where $p(x)$ is the probability density function of P and $r(x)$ is the probability density function of reference R. If $p(x)$ is exactly the same as $r(x)$, then $H(P||R){=}0$.

First, the derived state dynamics equation in (5.25) utilized in the proposed MFG opinion model are compared to the classical opinion models in (5.1) and (5.2). The purpose of the derived state dynamics equations is to be able to integrate opinion

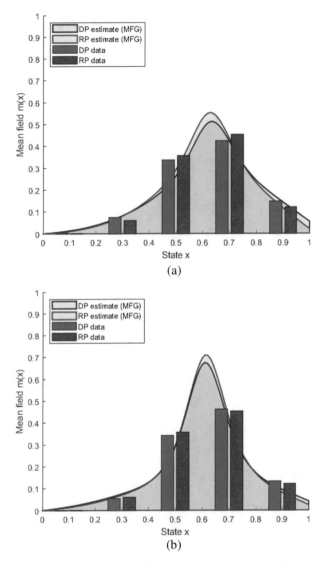

Fig. 5.5 Estimation of student opinion distribution using the proposed MFG opinion model. (**a**) April 2009. (**b**) June 2009

dynamics model such as the HK and FJ model in optimizations problems involving large number of players.

Figure 5.7 shows the relative entropy of the proposed MFG model with respect to the classical models. The size of the population and the average size of the neighborhood in terms of the fraction of the population have been varied. It can be concluded from the figure that as the size of the population increases, the relative entropy of the proposed MFG model decreases, as shown by the three

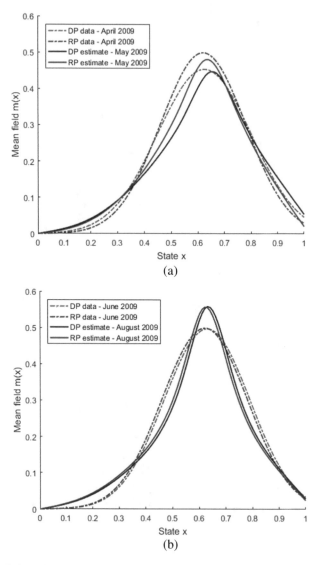

Fig. 5.6 Prediction of student opinion distribution using the proposed MFG opinion model. (**a**) May 2009. (**b**) August 2009

distinct curves in each figure. This signifies that the mean field of the opinions computed through the proposed MFG model closely approximates the mean field of the opinions computed through the classical models. Thus, the derived state dynamic equations becomes more valid for scenarios involving large number of participants. In addition, approximating the average neighborhood opinion (i.e., the classical models) using the average population opinion (i.e., the proposed MFG

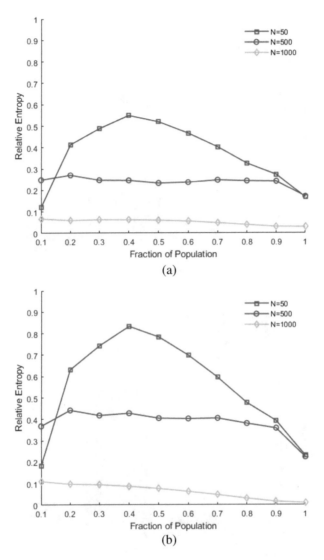

Fig. 5.7 Relative entropy of the proposed MFG opinion model with respect to the classical opinion models. (**a**) HK model. (**b**) FJ model

model) becomes more effective since the relative entropy is almost constant with varying neighborhood size (i.e., fraction of the population size).

Second, the performance of the estimation and prediction capability of the proposed MFG opinion model over the classical opinion models is explored. The classical models are utilized to estimate and predict the evolution of opinions. Then, the relative entropy of the proposed MFG and classical models with respect to the dataset are compared.

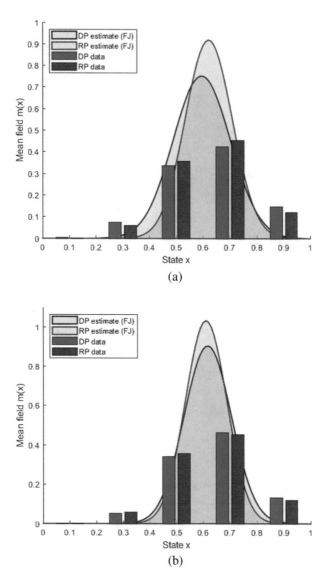

Fig. 5.8 Estimation of student opinion distribution using the classical opinion models. (**a**) April 2009. (**b**) June 2009

Back in Fig. 5.5, the mean field computed using the proposed MFG model has a relative entropy of 0.0633 for April and 0.1380 for June with respect to the data. Meanwhile, in Fig. 5.8, the mean field computed using the classical FJ model has a relative entropy of 0.3551 for April and 0.2919 for June with respect to the data. Based on these calculations, the data have been more accurately estimated by the proposed MFG opinion model than by the mean field computed from the classical model. The distribution of the opinions resulting from the classical models tend to concentrate around the mean more than the actual data. Meanwhile, the proposed MFG model can characterize not only the mean but also the other parts of the distribution more accurately. Thus, the relative entropy of the proposed MFG opinion model with respect to the data is smaller. For Fig. 5.6, the relative entropy of the predicted mean field for May and August cannot be computed since there is no available data for May and August to compare it to. However, the MFG model used to predict these mean fields is the same model that predicted the April and June mean fields which yield small relative entropy with respect to the data.

5.7 Related Works

Belief and opinion dynamics determine how social network users influence each other. Specifically, these models dictate how a user changes or updates its opinion based not only on its initial opinion but also on the opinions of the neighboring users. Well-known models of belief and opinion evolution are the DeGroot (DG) model [13], the Friedkin-Johnsen (FJ) model [3], and the Hegselmann-Krause (HK) model [1]. The DG model states that the current opinion of a user is the weighted sum of the opinions of the neighboring users, while the HK model states that the current opinion of a user is based on the average opinion of the neighboring users. Meanwhile, the FJ model builds on the DG model where the current opinion of a user is calculated by imposing a trade-off between the weighted sum of the neighboring opinions and the user's initial opinion. Variations of these models as well as other opinion dynamics models can be found in [14–17].

Since social networks are populated with large number of interacting users, mean field analysis has been utilized in a few works. In [18], the authors developed a model of social belief and opinion evolution. They derived a Fokker-Planck (FP) equation representing how the probability density of the users in the belief-personality space varies with time. Meanwhile, the authors of [19] proposed an MFG approach to model emulation, mimicry, and herding behaviors that can be observed when large number of social groups interact. This work considered social networks with large number of groups. An MFG was utilized in [20] to describe the propagation of opinion in social network according to a stochastic averaging process in the presence of an adversarial disturbance.

Other mathematical frameworks have been proposed as tools to model opinion dynamics in social networks. The authors of [21] proposed the evolution of opinions as well as their uncertainties in social networks as a fuzzy opinion network. The evolution of opinions on several interdependent topics as well as the convergence of mutually dependent opinions were investigated in [22]. Community detection, or finding connected users with similar opinion, was formulated as a multi-objective optimization problem through a graph K-means framework in [23]. A measurement analysis of user behavior in popular OSNs was implemented in [24]. The authors addressed two issues in OSNs: characterization of user activities and usage patterns and found out that these issues can be represented by well-known probability distributions.

The main differences of the proposed MPMFG-based work to the related works mentioned above are summarized as follows. First, this work focuses on the integration of the classical models (i.e., HK and FJ) to the MFG framework, in contrast with the non-MFG works. Consequently, the resulting MFG models aim to depict the opinion evolution in populations with large number of users where each user considers not only the opinion of its neighbors through the classical model but also its own behavior or preference through a cost function. Second, this work focuses on finite number of groups or populations and on the ability of a user to have an input or control on its own opinion. This is in contrast with other works focusing on large number of populations [19] and neighbor influence [25]. Third, this work focuses on the practical application of the proposed MFG model. Specifically, it takes advantage of the ability of the model to estimate and predict the evolution of opinion data from users. A social evolution dataset has been utilized to test the effectiveness of the proposed MFG models in opinion evolution. Finally, this work provides performance evaluations in order to show the validity of the proposed MFG models in estimating and predicting opinion evolution in a multiple-population social network.

5.8   Conclusion

Social networks are an important platform for social media users to express their opinion on social topics and events. Meanwhile, social network service providers aim to improve their service despite the growing demand from network users. One way of improving their service is to gain information and predict the behavior of the network users. Thus, belief and opinion evolution in social networks has gained the interest of researchers. Since the number of social network users has grown recently, an appropriate mathematical framework is necessary in order to analyze and model the behavior of social network users. Consequently, this work has proposed the application of a multiple-population mean field game (MPMFG) to gain knowledge about the belief and opinion evolution of social network users. Based on the simulations, the proposed MFG-based framework provides insights on how users and populations behave on a multiple-population social network. Using a social evolution dataset as a benchmark, the proposed MFG-based method allows for an effective approach to estimate and predict the distribution of opinions of social network users on a social topic.

References

1. R. Hegselmann, U. Krause, Opinion dynamics and bounded confidence models, analysis and simulations. J. Artif. Soc. Soc. Simul. **5**(3), 1–33 (2002)
2. G. Albi, L. Pareschi, M. Zanella, On the optimal control of opinion dynamics on evolving networks, in *System Modeling and Optimization*, vol. 494 (Springer, Berlin, 2017), pp. 58–67
3. N. Friedkin, E. Johnsen, Social influence networks and opinion change. Adv. Group Process. **16**(1), 1–29 (1999)
4. J.M. Lasry, P.L. Lions, Mean field games. Jpn. J. Math. **2**(1), 229–260 (2007)
5. D.A. Gomes, L. Nurbekyan, E.A. Pimentel, *Economic Models and Mean-field Game Theory*. 30° Colóquio Brasileiro de Matemática, Rio de Janeiro, July 2015
6. J.S. Rosenthal, *A First Look at Rigorous Probability Theory*, 2nd edn. (World Scientific, London, 2006)
7. A. Oppenheim, R.W. Schafer, *Discrete-Time Signal Processing* (Pearson Education, London, 2010)
8. A. Bensoussan, T. Huang, M. Lauriere, Mean field control and mean field game models with several populations. Optim. Control (2018). https://arxiv.org/abs/1810.00783
9. Y. Cao, S. Li, L. Petzold, R. Serban, Adjoint sensitivity analysis for differential-algebraic equations: the adjoint DAE system and its numerical solution. SIAM J. Sci. Comput. **23**(3), 1076–1089 (2003)
10. J.M. Schulte, Adjoint methods for Hamilton-Jacobi-Bellman equations. Diploma Thesis, University of Munster, Nov 2010
11. Y. Achdou, Finite difference methods for mean field games, in *Hamilton-Jacobi Equations: Approximations, Numerical Analysis and Applications* (Springer, Berlin, 2013)
12. A. Madan, M. Cebrian, S. Moturu, K. Farrahi, A. Pentland, Sensing the 'Health State' of a community. Pervasive Comput. **11**(4), 36–45 (2012)
13. M.H. DeGroot, Reaching a consensus. J. Am. Stat. Assoc. **69**(345), 118–121 (1974)
14. C. Chamley, A. Scaglione, L. Li, Models for the diffusion of beliefs in social networks. IEEE Signal Process. Mag. **30**(3), 16–29 (2018)
15. C. Castellano, S. Fortunato, V. Loreto, Statistical physics of social dynamics. Rev. Mod. Phys. **81**(2), 591–646 (2009)
16. A. Proskurnikova, R. Tempo, A tutorial on modeling and analysis of dynamic social networks, part I. Annu. Rev. Control **43**(1), 65–79 (2017)
17. H. Noorazar, K. Vixie, A. Talebanpour, Y. Hu, From classical to modern opinion dynamics. International J. Mod. Phys. C **31**(07), 2050101 (2020)
18. A. Nordio, A. Tarable, C.F. Chiasserini, E. Leonardi, Belief dynamics in social networks: a fluid-based analysis. IEEE Trans. Netw. Sci. Eng. **5**(4), 276–287 (2018)
19. D. Bauso, H. Tembine, T. Basar, Opinion dynamics in social networks through mean field games. SIAM J. Control. Optim. **54**(6), 3225–3257 (2016)
20. H. Tembine, D. Bauso, T. Basar, Robust linear quadratic mean-field games in crowd-seeking social networks, in *52nd IEEE Conference on Decision and Control, Florence*, Dec 2013, pp. 3134–3139
21. L.X. Wang, J.M. Mendel, Fuzzy opinion networks: a mathematical framework for the evolution of opinions and their uncertainties across social networks. IEEE Trans. Fuzzy Syst. **24**(4), 880–905 (2016)
22. S.E. Parsegov, A.V. Proskurnikov, R. Tempo, N.E. Friedkin, Novel multidimensional models of opinion dynamics in social networks. IEEE Trans. Autom. Control **62**(5), 2270–2285 (2017)
23. Z. Bu, H.J. Li, C. Zhang, J. Cao, A. Li, Y. Shi, Graph K-means based on leader identification, dynamic game and opinion dynamics. IEEE Trans. Knowl. Data Eng. **32**(7), 1348–1361 (2019)
24. L. Gyarmati, T.A. Trinh, Measuring user behavior in online social networks. IEEE Netw. **24**(5), 26–31 (2010)
25. D. Bauso, R. Pesenti, M. Tolotti, Opinion dynamics and stubbornness via multi-population mean field games. Springer J. Optim. Theory Appl. **170**(1), 266–293 (2016)

Chapter 6
Mean-Field-Type Game for Multi-Access Edge Computing Networks

Multi-access edge computing (MEC) has been proposed to reduce the latency inherent in traditional cloud computing. One of the services offered in an MEC network (MECN) is computation offloading in which computing nodes, with limited capabilities and performance, can offload computation-intensive tasks to other computing nodes in the network. Meanwhile, mean-field-type game (MFTG) has been recently applied in engineering applications where the number of decision makers is finite and a decision maker can be distinguishable from other decision makers and have a non-negligible effect on the total utility of the network. Since MECNs are implemented through a finite number of computing nodes and the computing capability of a computing node can affect the state (i.e., the number of computation tasks) of the network, non-cooperative and cooperative MFTG approaches are proposed in this work to formulate the computation offloading problems in an MECN. In these scenarios, the goal of each computing node is to offload a portion of the aggregate computation tasks from the network that minimizes a specific cost. Then, a direct approach is utilized to calculate the optimal computation offloading solution (i.e., optimal portion of the aggregate computation tasks) of a computing node. Furthermore, non-cooperative and cooperative MFTG-based algorithms are proposed to implement computation offloading in an MECN. Finally, simulations are presented to show the significance and advantage of the proposed MFTG-based computation offloading algorithms over the traditional methods.

6.1 System Model

Figure 6.1 shows the system model proposed in this research. The end user devices (EUs) offload computation-intensive tasks that cannot be performed locally to the task aggregator (TA) in the area or cell the EU is located. Each EU decides to

R. A. Banez et al., *Mean Field Game and its Applications in Wireless Networks*,
Wireless Networks, https://doi.org/10.1007/978-3-030-86905-2_6

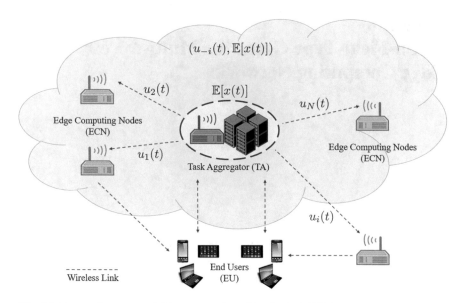

Fig. 6.1 Proposed system model for computation offloading in MECNs

offload based on algorithms presented in the literature such as in [1]. Then, the TA combines all the computation tasks submitted by the EUs in the area. It organizes the computation tasks to reduce redundancy and overloading of computation tasks. Moreover, it performs a portion of the tasks and directly sends the results to the corresponding EUs. Afterwards, the TA offloads parts of the remaining aggregate computation tasks to the edge computing nodes (ECNs) in the cell. Each ECN is capable of performing computation-intensive tasks and is more powerful than a typical mobile EU equipment. For instance, a typical ECN has a computing power of about 10,000 to 100,000 times that of a mobile phone [2]. After the ECNs perform their respective offloaded computation tasks, they transmit the results to the requesting EUs.

Consider a cell of an MECN consisting of one TA and a finite set $\mathcal{N}$ of ECNs with $|\mathcal{N}| = N$. The time horizon defined as $t \in [0, T]$ is finite, where $T > 0$ is the terminal time. Let the network state $x(t)$ be the number of aggregate computation tasks to be offloaded by the TA to the ECNs at time t. In addition, denote the network state dynamics $x'(t) = dx(t)/dt$ as the change or evolution of the number of aggregate computation tasks with respect to time. Also, let the admissible computation offloading control $u_i(t)$ be the portion of $x(t)$ offloaded by ECN i from the TA at time t. The goal of each ECN $i \in \mathcal{N}$ is to determine its optimal control $u_i^*(t)$ that minimizes its cost, defined by a cost function J_i, subject to network state dynamics $x'(t)$.

In the following subsections, the cost function and network state dynamics equation are presented. Important parameters that influence the optimal control of an ECN are also introduced.

6.1.1 Cost Functions

In this work, the ECNs are assumed to follow a quadratic cost function because of its desirable economic properties such as monotonicity, concavity, non-decreasing [3].

Let the consumed energy of ECN i per CPU cycle be $\varepsilon_i = \kappa_{e,i} f_i^2$ [4], where $\kappa_{e,i}$ is a constant depending on the architecture of the CPU of ECN i and f_i is the computing capability (i.e., the number of CPU cycles per unit time) of ECN i. To calculate the cost associated with the energy consumed by a certain number of computation tasks, define the energy cost coefficient e_i as the cost (per unit time) of energy spent by computing node i per squared number of CPU cycles,

$$e_i = w_{e,i} \varepsilon_i^2 = w_{e,i} (\kappa_{e,i} f_i^2)^2, \tag{6.1}$$

where the constant $w_{e,i}$ is the weight or significance assigned to energy consumption cost. A higher value of $w_{e,i}$ means that an ECN prioritizes minimizing its energy consumption.

Meanwhile, to calculate the cost corresponding to the execution or computation time of a certain number of computation task, define the computation time cost coefficient τ_i as the cost (per unit time) associated with execution time spent by ECN i per squared number of CPU cycles,

$$\tau_i = \frac{w_{d,i}}{f_i^2}, \tag{6.2}$$

where the constant $w_{d,i}$ is the weight assigned by ECN i to the computation time cost. A higher value of $w_{d,i}$ means that ECN i prioritizes minimizing the cost from computation time.

Lastly, to quantify the cost earned by the TA from offloading computation task to ECN i, define the offloading cost coefficient ρ_i as the cost (per unit time) incurred by the TA per squared number of aggregate computation task. It takes into account the cost not associated with computation by an ECN i such as processing and transmission from the TA to ECN i.

Combining these cost coefficients with the network state $x(t)$ and the control $u_i(t)$ yields the running cost function that tells how much the cost increases or decreases with time,

$$L_i(x(t), u_i(t), t) = \frac{1}{2}[\rho_i x^2(t) + (\tau_i + e_i) u_i^2(t)]. \tag{6.3}$$

Since the goal of computation offloading is to offload tasks from the TA to the ECNs, the number of computation tasks $x(t)$ that remains at the TA at terminal time T is penalized. That is, $x(T)$ is considered as a part of the cost to be minimized. Since the cost at terminal time T is proportional to the number of computation tasks $x(T)$, the terminal cost function for ECN i is stated as

$$\Phi_i(x(T), T) = \frac{1}{2}\rho_i x^2(T). \tag{6.4}$$

In other words, it computes the cost incurred by ECN i based on the network state $x(t)$ at $t = T$.

6.1.2   Network State Dynamics Equation

The network state dynamics $x'(t)$ refers to the evolution of the network state $x(t)$ with respect to time t. In the computation offloading system model, $x'(t)$ refers to the dynamics or change in the number of aggregate computation tasks at the TA with time. Let $q_{in}(t)$ be the incoming rate of the computation tasks to the TA. Then, the number of computation tasks $x(t)$ at the TA is related, $q_{in}(t) = r_0 x(t)$, with r_0 defined as

$$r_0 = \frac{R_0}{C_0} = \frac{1}{C_0} \sum_{j=1}^{M} B_j \log_2(1 + \gamma_j) = \frac{1}{C_0} \sum_{j=1}^{M} B_j \left(1 + \frac{P_j g_{0,j}}{N_0 + I_j}\right), \tag{6.5}$$

where R_0 is the maximum incoming rate of computation task the TA handle and C_0 is the capacity or the maximum number of computation tasks the TA can store. Physically, r_0 is the frequency at which computation tasks arrive at the TA. Meanwhile, R_0 is the sum of the rates the TA receives from M EUs, B_j is the channel bandwidth for EU j, and γ_j is the signal-to-interference-plus-noise (SINR) ratio between the TA and EU j, where P_j refers to EU j transmit power, $g_{0,j}$ is the channel gain between the TA and EU j, N_0 is the background noise power, and I_j is the interference noise power experience by EU j.

On the other hand, the outgoing rate $q_{out}(t)$ of computation task from the TA is affected by the computation offloading control $u_i(t)$ of ECN $i \in \mathcal{N}$. Hence, $q_{out}(t) = \sum_{i=1}^{N} r_i u_i(t)$ with

$$r_i = \frac{R_i}{C_i} = \frac{B_i \log_2(1 + \gamma_i)}{C_i} = \frac{B_i}{C_i} \log_2 \left(1 + \frac{P_i g_{i,0}}{N_0 + I_i}\right), \tag{6.6}$$

where R_i is the maximum outgoing rate of computation task to ECN i, C_i is the capacity or the maximum number of computation tasks ECN i can handle, and consequently, r_i is the frequency at which computation tasks arrive at ECN i. In addition, B_i is the channel bandwidth of ECN i and γ_i is the SINR between the TA and ECN i, where P_i is the transmit power of ECN i, $g_{i,0}$ is the channel gain between the TA and ECN i, N_0 is the background noise power, and I_i is the interference power experienced by ECN i.

Since the total rate of computation task $x'(t) = q_{in}(t) - q_{out}(t)$, then the network state dynamics equation can be written as

$$dx(t) = \left(r_0 x(t) - \sum_{i=1}^{N} r_i u_i(t) \right) dt, \tag{6.7}$$

which is similar in form with the state dynamics equation used in [5].

To summarize, the cost function is affected by $u_i(t)$ since the cost depends on the number of tasks ECN i offloads from the TA. On the other hand, the state dynamic equation is affected by $r_i u_i(t)$ since the change in the number of $x(t)$ depends on the rate $r_i u_i(t)$ at which tasks are offloaded to ECN i. Moreover, the linear state dynamic equation in (6.7) can also represent a state dynamic equation of the form $dx(t) = f(t) dt$ through linearization at sampling time $t = t_n$ of (1) the function $F(t) = \int_0^t f(s) ds$,

$$F(t) = F(t_n) + \frac{\partial F}{\partial t}|_{t=t_n} \cdot (t - t_n),$$

and (2) the solution $x(t)$ of (6.7),

$$x(t) = x(t_n) + \frac{\partial x}{\partial t}|_{t=t_n} \cdot (t - t_n).$$

Consequently, $f(t_n) = \frac{\partial F}{\partial t}|_{t=t_n} = \frac{\partial x}{\partial t}|_{t=t_n}$. For instance, if $x(t) = x_0 e^{g(t)}$, then $f(t_n) = \frac{\partial g}{\partial t}|_{t=t_n} \cdot x(t_n)$. If $x(t)$ follows a distribution function such as a Poisson process, then the state dynamic equation $dx(t)$ can model the transition between the sampling times.

In the next section, these formulations are extended to be able to adapt an MFTG approach. The main feature of this method is the addition of mean field terms $\bar{x}(t)$ and $\bar{u}(t)$ in the cost functions and the state dynamics equation. Consequently, each ECN now aims to compute its optimal control $u_i^*(t)$ given the mean field terms $\bar{x}(t)$ and $\bar{u}(t)$ and the differences $x(t) - \bar{x}(t)$ and $u(t) - \bar{u}(t)$.

6.2   Mean-Field-Type Game Problem Formulation

The theory of MFG, introduced in [6–9], has been used in a variety of applications formulated as games among a large number of decision makers that aim to optimize their own payoffs or cost functions subject to a state dynamic equation. The main concept behind MFG is that each decision maker determines its optimal strategy (i.e., the strategy or action that optimizes its payoff or cost function) based on an aggregate information about the states of other decision makers. In other words, a decision maker computes its optimal strategy based on a statistical distribution of the states of other decision makers (i.e., a mean-field term) instead on a full knowledge of the states of other decision makers.

According to [10], most MFG models share the following assumptions: (1) there are infinitely many decision makers, (2) the decision makers are indistinguishable, and (3) a decision maker has negligible effect on the global utility. However, in engineering applications, these assumptions may be difficult to prove. Consequently, a more relaxed MFTG has been proposed in the literature. In MFTG, the number of decision makers may be infinite or finite, the decision makers may not be indistinguishable, and finally, a decision maker may have a significant effect on the global utility. Applications of MFTG include distributed power networks [11], network security [12], and multilevel building evacuation [13].

In this section, computation offloading in MECN is formulated as an MFTG. First, the cost functions to be minimized by an ECN are derived. Then, the state dynamics, the differential equation constraint of the minimization problem, is formulated. The MFTG cost functions and state dynamic equation contain mean field terms that quantify the behavior or strategy of all the computing nodes. These terms are added to the cost function so that each computing node can minimize the variance in the network state as well as the variance in computing node control. In the last subsection, the two MFTG computation offloading problems are stated, a non-cooperative MFTG problem where the ECNs minimize their own cost function independently and a cooperative MFTG problem where the ECNs minimize a single global cost function.

6.2.1 Preliminaries

The network state or state of the TA $x(t)$ refers to the number of aggregate computation tasks to be offloaded to the ECNs. An admissible computation offloading control or strategy $u_i(t)$ of ECN i refers to a portion of $x(t)$ it can offload from the TA, while the set $\mathcal{U}_i$ denotes the set of all admissible controls of ECN i. Vector $u(t) = [u_i(t)]_{i \in \mathcal{N}}$ contains the control of all the ECNs in the cell, while the vector $u_{-i}(t) = [u_i(t)]_{i \in \mathcal{N} \setminus i}$ contains the control of all ECNs in the cell except ECN i.

The following subsections present the cost functions and state dynamic equation in an MFTG setting. The main difference in the formulations to follow is the inclusion of mean field terms $\bar{x}(t) = \mathbb{E}[x(t)]$ and $\bar{u}(t) = \mathbb{E}[u(t)]$. Consequently, a tilde $\sim$ is put on top of the MFTG cost function $\tilde{J}_i$ and state dynamic function $\tilde{f}$ to differentiate them from their mean-field-free counterparts. Afterwards, the resulting MFTG-based computation offloading problems are stated.

6.2.2 Cost Functions

The total cost function $\tilde{J}_i(u)$ of ECN i consists of the running cost function $\tilde{L}_i$, which corresponds to the accumulated cost of ECN i for performing a portion $u_i(t)$ of $x(t)$, and the terminal cost function $\tilde{\Phi}_i$, which penalizes the computing node at terminal time $t = T$ depending on how far the network state $x(t)$ is from a target state (e.g., $x(t) = 0$, when all of the aggregate computation tasks are offloaded). Mathematically,

$$\tilde{J}_i(u) = \mathbb{E}\left[\int_0^T \tilde{L}_i(x, u, \bar{x}, \bar{u}, t)\, dt + \tilde{\Phi}_i(x, \bar{x}, T)\right]. \tag{6.8}$$

However, the running cost $\tilde{L}_i(x, u, \bar{x}, \bar{u}, t)$ for MFTG differs from that in (6.3) since $\tilde{L}_i$ depends on the expected values of the network state $\bar{x}(t)$ and the control $\bar{u}$. The expected values have been included in the cost function because these values are assumed to be known, and consequently, the difference to these expected values, $x(t) - \bar{x}(t)$ and $u_i(t) - \bar{u}_i(t)$. Hence, $\mathbb{E}\left[\tilde{L}_i(x, u, \bar{x}, \bar{u}, t)\right]$ is given by

$$\mathbb{E}\left[\tilde{L}_i(x, u, \bar{x}, \bar{u}, t)\right] = \frac{1}{2}\mathbb{E}\left[\rho_i x^2(t) + \bar{\rho}_i \bar{x}^2(t) + (\tau_i + e_i)u_i^2(t) + (\bar{\tau}_i + \bar{e}_i)\bar{u}_i^2(t)\right],$$

$$= \frac{1}{2}\mathbb{E}\left[\rho_i(x(t) - \bar{x}(t))^2 + (\rho_i + \bar{\rho}_i)\bar{x}^2(t) + (\tau_i + e_i)(u_i(t) - \bar{u}_i(t))^2\right.$$

$$\left. + (\tau_i + \bar{\tau}_i + e_i + \bar{e}_i)\bar{u}_i^2(t)\right],$$

$$= \frac{1}{2}\left(\rho_i \mathrm{var}[x(t)] + (\rho_i + \bar{\rho}_i)\bar{x}^2(t) + (\tau_i + e_i)\mathrm{var}[u_i(t)]\right.$$

$$\left. + (\tau_i + \bar{\tau}_i + e_i + \bar{e}_i)\bar{u}_i^2(t)\right), \tag{6.9}$$

where $\bar{\rho}_i$, $\bar{\tau}_i$ and $\bar{e}_i$ refer to the mean of cost coefficients defined in Sect. 6.1.1.

Similarly, the terminal cost $\tilde{\Phi}_i(x, \bar{x}, T)$ depends as well on the expected value of the network state at time T so that the goal of an ECN i is to minimize $\mathrm{var}[x(t)]$

$$\mathbb{E}\left[\tilde{\Phi}_i(x, \bar{x}, T)\right] = \frac{1}{2}\mathbb{E}\left[\rho_i x^2(T) + \bar{\rho}_i \bar{x}^2(T)\right],$$

$$= \frac{1}{2}\mathbb{E}\left[\rho_i(x(T) - \bar{x}(T))^2 + (\rho_i + \bar{\rho}_i)\bar{x}^2(T)\right], \tag{6.10}$$

$$= \frac{1}{2}\left(\rho_i \mathrm{var}[x(T)] + (\rho_i + \bar{\rho}_i)\bar{x}^2(T)\right).$$

The quadratic cost functions presented in this work refer to the penalty incurred by the network through the ECNs for executing a specific number of computation tasks. While the ECNs are constrained in terms of their local energy consumption and execution time through e_i and τ_i, respectively, the penalty allows the ECNs follow a network-wide algorithm that optimizes the network performance. Hence, the physical meaning of the costs refers to the penalty set by the network to the ECNs. These penalties are based on physical quantities the computing nodes spend when performing computation tasks. The number of computation tasks offloaded by each ECN is limited by these penalties. Moreover, the costs are also indicators of network performance since a low cost may indicate a low terminal cost $\tilde{\Phi}_i(x, \bar{x}, T)$

which means a low number of un-offloaded tasks remain in the TA; also, a low cost may indicate a low running cost $\tilde{L}_i(x, u, \bar{x}, \bar{u}, t)$ which means the ECNs execute the number of computation tasks that satisfy their local energy and time constraints.

6.2.3   Network State Dynamics Equation

The network state dynamics $x'(t)$ of refers to the evolution of the state $x(t)$ of the TA with respect to time t. In (6.7), the network state dynamics is affected by the current network state $x(t)$ and the controls $u_i(t)$, $\forall i \in \mathcal{N}$. Like in the MFTG cost functions, the expected values $\bar{x}$ and $\bar{u}$ are included in the MFTG state dynamic equation $x'(t)$ so that the evolution of the state can be formulated in terms of the deviation of $x(t)$ and $u(t)$ from their respective expected values. Consequently, the updated network state dynamics of a computation offloading in an MFTG setting is

$$dx(t) = \tilde{f}(x, u, \bar{x}, \bar{u}) \, dt + \sigma \, dW(t), \tag{6.11}$$

where $W(t)$ is a standard Wiener process, σ is a coefficient that captures the randomness in the state dynamics, the drift term $\tilde{f}(x, u, \bar{x}, \bar{u}, t)$ is given by

$$\tilde{f}(x, u, \bar{x}, \bar{u}, t) = r_0 x(t) + \bar{r}_0 \bar{x}(t) - \left(\sum_{i=1}^{N} r_i u_i(t) + \sum_{i=1}^{N} \bar{r}_i \bar{u}_i(t) \right), \tag{6.12}$$

and the coefficients are defined as $\bar{r}_0 = \mathbb{E}[\frac{R_0}{C_0}]$ and $\bar{r}_i = \mathbb{E}[\frac{R_i}{C_i}] = \mathbb{E}[\frac{B_i \log_2(1+\gamma_i)}{C_i}]$. The drift term can be written in an equivalent form

$$\tilde{f}(x, u, \bar{x}, \bar{u}, t) = r_0(x(t) - \bar{x}(t)) + (r_0 + \bar{r}_0)\bar{x}(t)$$

$$- \left(\sum_{i=1}^{N} r_i(u_i(t) - \bar{u}_i(t)) + \sum_{i=1}^{N} (r_i + \bar{r}_i)\bar{u}_i(t) \right), \tag{6.13}$$

which expresses the network state dynamics as the sum of the mean fields $\bar{x}(t)$ and $\bar{u}_i(t)$ and the terms $x(t) - \bar{x}(t)$ and $u_i(t) - \bar{u}_i(t)$. In the following computation offloading scenarios, since the mean number of computation tasks $x(t)$ is tracked, the difference $x(t) - \bar{x}(t)$ of the current state $x(t)$ from mean state $\bar{x}(t)$ can be easily known. The same principle is applied with $u_i(t)$ and $\bar{u}_i(t)$.

6.2.4 Non-cooperative Problem

Consider an MECN consisting of $N \geq 2$ ECNs. Each ECN is capable of offloading and performing computation tasks from the TA. In addition, suppose the MECN implements a non-cooperative scenario where each ECN computes its offloading strategy by minimizing its own cost function. If the cost function of ECN i is defined by (6.8), then in a non-cooperative setting each ECN i tries to solve the MFTG problem

$$
\inf_{u_i \in \mathscr{U}_i} \quad \tilde{J}_i(u) = \frac{1}{2}\mathbb{E}\bigg[\int_0^T \Big[\rho_i(x(t) - \bar{x}(t))^2 + (\rho_i + \bar{\rho}_i)\bar{x}^2(t)
$$

$$
+ (\tau_i + e_i)(u_i(t) - \bar{u}_i(t))^2 + (\tau_i + \bar{\tau}_i + e_i + \bar{e}_i)\bar{u}_i^2(t) \Big] dt
$$

$$
+ \rho_i(x(T) - \bar{x}(T))^2 + (\rho_i + \bar{\rho}_i)\bar{x}^2(T) \bigg],
$$

$$
\text{subject to} \quad dx(t) = \bigg[r_0(x(t) - \bar{x}(t)) + (r_0 + \bar{r}_0)\bar{x}(t) - \bigg(\sum_{i=1}^N r_i(u_i(t) - \bar{u}_i(t))
$$

$$
+ \sum_{i=1}^N (r_i + \bar{r}_i)\bar{u}_i(t) \bigg) \bigg] dt + \sigma(t)\, dW(t),
$$

$$
x(0) = x_0,
$$

(6.14)

where $\bar{x}(t) < +\infty$. Any control $u_i^*(t)$ that satisfies (6.14) is the best-response of computing node i to $(u_{-i}, \mathbb{E}[x(t)])$.

Definition 6.1 Any control $u_i^*(t) \in \mathscr{U}_i$ satisfying (6.14) is called a risk-neutral best-response control of computing node i to the control $u_{-i} \in \Pi_{j \in \mathscr{N}}\mathscr{U}_j$ of the other computing nodes $j \neq i$.

The set of best-response controls of computing node i is defined by $\mathscr{BR}_i$: $\Pi_{j \in \mathscr{N}}\mathscr{U}_j \rightarrow 2^{\mathscr{U}_i}$, where $2^{\mathscr{U}_i}$ is the set of subsets of $\mathscr{U}_i$. Using the concept of best-response control strategy, a Nash equilibrium of (6.14) is (u_i^*, u_{-i}^*), where every ECN i solves their best-response control u_i^*.

Definition 6.2 A Nash equilibrium of the mean-field-type game in (6.14) is a control profile (u_i^*, u_{-i}^*), such that for every computing node i,

$$
\tilde{J}_i(u_i^*, u_{-i}^*) \leq \tilde{J}_i(u_i, u_{-i}^*), \forall u_i \in \mathscr{U}_i.
$$

(6.15)

6.2.5   Cooperative Problem

Suppose the ECNs try to jointly minimize a single global cost function $\tilde{J}_0(u) = \mathbb{E}[\sum_{i=1}^{N} \tilde{J}_i(u)]$ where $u = (u_1, \ldots, u_N)$ is the computation offloading control profile in a cooperative setting. Then, the corresponding cooperative MFTG problem is given by

$$\inf_{u_i \in \mathscr{U}_i} \quad \tilde{J}_0(u) = \frac{1}{2}\mathbb{E}\left[\sum_{i=1}^{N} \int_0^T \left[\rho_i(x(t) - \bar{x}(t))^2 + (\rho_i + \bar{\rho}_i)\bar{x}^2(t) \right. \right.$$

$$+ (\tau_i + e_i)(u_i(t) - \bar{u}_i(t))^2 + (\tau_i + \bar{\tau}_i + e_i + \bar{e}_i)\bar{u}_i^2(t) \Big] dt$$

$$+ \rho_i(x(T) - \bar{x}(T))^2 + (\rho_i + \bar{\rho}_i)\bar{x}^2(T) \Big],$$

$$\text{subject to} \quad dx(t) = \left[r_0(x(t) - \bar{x}(t)) + (r_0 + \bar{r}_0)\bar{x}(t) - \left(\sum_{i=1}^{N} r_i(u_i(t) - \bar{u}_i(t)) \right. \right.$$

$$+ \sum_{i=1}^{N}(r_i + \bar{r}_i)\bar{u}_i(t) \bigg) \bigg] dt + \sigma(t)\, dW(t),$$

$$x(0) = x_0.$$

(6.16)

Any control profile $u^* = (u_1^*, \ldots, u_N^*)$ that satisfies (6.16) is a global optimum control profile that minimizes the global cost function $\tilde{J}_0$.

The next section provides the method proposed in [14] to solve for a solution of linear-quadratic MFTGs such as (6.14) and (6.16).

6.3   Linear-Quadratic Mean-Field-Type-Game Solution Using a Direct Method

The MFTG problems defined in (6.14) and (6.16) are called linear-quadratic MFTGs (LQMFTG) since the cost functional is quadratic and the state dynamics is linear with respect to the state and control. Because of their special form, the authors in [14] proposed a direct approach in computing the optimal control $u_i^*(t)$ of LQMFTG. The proposed method can solve an LQMFTG without solving coupled partial differential equations. The authors proved that the proposed direct approach to LQMFTG yields the same solution as an analytical approach. Based on this method, this section presents the main concepts in deriving the solution for the non-cooperative and cooperative MFTG problems introduced in the previous section. The solution $u_i^*(t)$ to each problem refers to the computation offloading control or the number of computation tasks $u_i^*(t)$ ECN i must offload from the TA in order to minimize the corresponding cost of the problem. In other words, the optimal control is the number of computation tasks to be offloaded by an ECN such that the penalty incurred by the network due to the number of executed computation tasks by the ECN and the remaining tasks at the TA are minimized.

6.3.1 Non-cooperative Solution

The direct method for the LQMFTG problem starts with choosing a guess cost functional $\phi_i(x, t)$. Since the cost functional J_i is quadratic, the corresponding $\phi_i(x, t)$ is quadratic as well,

$$\phi_i(x, t) = \frac{1}{2}\alpha_i(x - \bar{x})^2 + \frac{1}{2}\beta_i\bar{x}^2 + \gamma_i\bar{x} + \delta_i,$$

where α_i, β_i, γ_i, and δ_i are restricted to time-invariant coefficients for $[0, T]$.

Then, apply the Ito's formula for a drift-diffusion process to $\phi_i(x, t)$ with $t = T$,

$$\phi_i(x(T), T) = \phi_i(x(0), 0) + \int_0^T \left(\partial_t\phi_i + \tilde{f}(x, u, \bar{x}, \bar{u}, t)\partial_x\phi_i + \frac{\sigma^2}{2}\partial_{xx}\phi_i\right)dt$$

$$+ \int_0^T \sigma(t)\partial_x\phi_i\, dW(t).$$

(6.17)

The next step is to compute and substitute the partial derivatives $\partial_t\phi_i$, $\partial_x\phi_i$, and $\partial_{xx}\phi_i$ to $\phi_i(x(T), T)$ and take its expectation, $\mathbb{E}[\phi_i(x(T), T) - \phi_i(x(0), 0)]$. Afterwards, the gap $\tilde{J}_i - \mathbb{E}[\phi_i(x(0), 0)]$ is calculated.

Finally, the optimal control u_i^* is derived from $\min_{u_i \in \mathscr{U}_i} \tilde{J}_i(u)$ using the appropriate optimality principles. A control u_i is called a feedback control if it is a function of time t and the state $x(t)$. To compute the best-response control u_i^* of computing node i to feedback strategies u_j, $j \neq i$, complete the square of the gap $\tilde{J}_i - \mathbb{E}[\phi_i(x(0), 0)]$,

$$\tilde{J}_i - \mathbb{E}[\phi_i(x(0), 0)] = \frac{1}{2}\mathbb{E}\left[\int_0^T (\tau_i + e_i)\left(u_i - \bar{u}_i - \frac{r_i}{\tau_i + e_i}\alpha_i(x - \bar{x})\right)^2 dt\right]$$

$$+ \frac{1}{2}\mathbb{E}\left[\int_0^T (\tau_i + \bar{\tau}_i + e_i + \bar{e}_i)\left(\bar{u}_i - \beta_i\frac{r_i + \bar{r}_i}{\tau_i + \bar{\tau}_i + e_i + \bar{e}_i}\bar{x}\right)^2 dt\right]$$

$$+ \frac{1}{2}\mathbb{E}\left[\int_0^T \sigma^2\alpha_i\, dt\right].$$

(6.18)

Consequently, the equivalent objective functional becomes

$$\inf_{u_i \in \mathcal{U}_i} \tilde{J}_i = \frac{1}{2}\alpha_i(0)\text{var}[x(0)] + \frac{1}{2}\beta_i(0)(\mathbb{E}[x(0)])^2 + \frac{1}{2}\mathbb{E}\left[\int_0^T \sigma^2(t)\alpha_i(t)\,dt\right].$$
(6.19)

Using this equivalent objective functional, the following theorem holds for the optimal control u_i^*.

Theorem 6.1 *Let the cost functional $\tilde{J}_i(u)$ of an LQMFTG problem take the form $\phi_i(x,t) = \frac{1}{2}\alpha_i(x-\bar{x})^2 + \frac{1}{2}\beta_i\bar{x}^2 + \gamma_i\bar{x} + \delta_i$, where α_i, β_i, γ_i, and δ_i are constants. Then, the optimal control $u_i^*(t)$ associated with the problem is given by*

$$u_i^*(t) = \frac{r_i}{\tau_i + e_i}\alpha_i(x-\bar{x}) + \frac{r_i + \bar{r}_i}{\tau_i + \bar{\tau}_i + e_i + \bar{e}_i}\beta_i\bar{x},$$
(6.20)

where the constants α_i and β_i solve the following equations, respectively,

$$\frac{r_i^2}{\tau_i + e_i}\alpha_i^2 + 2\left(\sum_{j=1,j\neq i}^N \frac{r_j^2}{\tau_j + e_j}\alpha_j - r_0\right)\alpha_i - \rho_i = 0,$$

$$\frac{(r_i + \bar{r}_i)^2}{\tau_i + \bar{\tau}_i + e_i + \bar{e}_i}\beta_i^2 + 2\left(\sum_{j=1,j\neq i}^N \frac{(r_j + \bar{r}_j)^2}{\tau_j + \bar{\tau}_j + e_j + \bar{e}_j}\beta_j\right.$$
(6.21)

$$\left.-(r_0 + \bar{r}_0)\right)\beta_i - (\rho_i + \bar{\rho}_i) = 0,$$

and the mean field term $\bar{x}(t)$ is given by

$$\bar{x}(t) = \bar{x}(0)\exp\left[\int_0^t \left((r_0 + \bar{r}_0) - \sum_{i=1}^N \frac{\beta_i(r_i + \bar{r}_i)^2}{\tau_i + \bar{\tau}_i + e_i + \bar{e}_i}\right)ds\right],$$
(6.22)

and $\bar{u}_i$ has been expressed as $\beta_i(r_i + \bar{r}_i)/(\tau_i + \bar{\tau}_i + e_i + \bar{e}_i)\bar{x}$.

Proof The optimal control u_i^* is obtained by minimizing the following terms with respect to control u_i and $\bar{u}_i$,

$$\frac{\partial}{\partial u_i}\left[(\tau_i + e_i)\left(u_i - \bar{u}_i - \frac{r_i}{\tau_i + e_i}\alpha_i(x-\bar{x})\right)^2\right.$$

$$\left. + (\tau_i + \bar{\tau}_i + e_i + \bar{e}_i)\left(\bar{u}_i - \beta_i\frac{r_i + \bar{r}_i}{\tau_i + \bar{\tau}_i + e_i + \bar{e}_i}\bar{x}\right)^2\right] = 0,$$

which yields $u_i = \frac{r_i}{\tau_i + e_i}\alpha_i(x-\bar{x}) + \bar{u}_i$, where $\bar{u}_i = \beta_i(r_i + \bar{r}_i)/(\tau_i + \bar{\tau}_i + e_i + \bar{e}_i)\bar{x}$. Meanwhile, the mean field $\bar{x}(t)$ is derived by taking the expectation of the state dynamic equation in (6.14) and then solving the resulting differential equation for $\bar{x}(t)$.

Theorem 6.1 states that the optimal number of computation tasks ECN i must offload from the TA in a non-cooperative scenario is given in (6.20). This number minimizes the cost incurred by ECN i where α_i and β_i satisfy the conditions (6.21), and the mean field $\bar{x}(t)$ satisfies (6.22). It can be deduced that $u_i^*(t)$ not only depends on $x(t)$ but also on how much $x(t)$ exceeds $\bar{x}(t)$. Moreover, α_i and β_i reflect the weights of how much $u_i^*(t)$ depends on $x(t) - \bar{x}(t)$ and $\bar{x}(t)$, respectively.

6.3.2   Cooperative Solution

To obtain the global optimum solution to the cooperative LQMFTG problem in (6.16), the procedure stated in the previous subsection is followed. Hence, the LQMFTG problem in (6.16) is equivalent to

$$\inf_{u_1,\ldots,u_N} \tilde{J}_0 = \frac{1}{2}\alpha_0(0)\mathrm{var}[x(0)] + \frac{1}{2}\beta_0(0)(\mathbb{E}[x(0)])^2 + \frac{1}{2}\mathbb{E}\left[\int_0^T \sigma^2(t)\alpha_0(t)\,dt\right].$$

(6.23)

The corresponding optimal control $u_i^*(t)$ is given by the following theorem.

Theorem 6.2 *Let the cost functional $\tilde{J}_0(u)$ of an LQMFTG problem take the form $\phi_0(x,t) = \frac{1}{2}\alpha_0(x-\bar{x})^2 + \frac{1}{2}\beta_0\bar{x}^2$, where α_0 and β_0, are constants. Then, the optimal control $u_i^*(t)$ associated with the problem is given by*

$$u_i^*(t) = \frac{r_i}{\tau_i + e_i}\alpha_0(x-\bar{x}) + \frac{r_i + \bar{r}_i}{\tau_i + \bar{\tau}_i + e_i + \bar{e}_i}\beta_0\bar{x},$$

(6.24)

where the constants α_0 and β_0 solve the following equations, respectively,

$$\left(\sum_{i=1}^N \frac{r_i^2}{\tau_i + e_i}\right)\alpha_0^2 - 2r_0\alpha_0 - \rho_0 = 0,$$

$$\left(\sum_{i=1}^N \frac{(r_i + \bar{r}_i)^2}{\tau_i + \bar{\tau}_i + e_i + \bar{e}_i}\right)\beta_0^2 - 2(r_0 + \bar{r}_0)\beta_0 - (\rho_0 + \bar{\rho}_0) = 0,$$

(6.25)

and the mean field term $\bar{x}(t)$ is given by

$$\bar{x}(t) = \bar{x}(0)\exp\left[\int_0^t \left((r_0 + \bar{r}_0) - \beta_0 \sum_{i=1}^N \frac{(r_i + \bar{r}_i)^2}{\tau_i + \bar{\tau}_i + e_i + \bar{e}_i}\right)ds\right],$$

(6.26)

and $\bar{u}_i$ has been expressed as $\beta_0(r_i + \bar{r}_i)/(\tau_i + \bar{\tau}_i + e_i + \bar{e}_i)\bar{x}$.

Proof The optimal control u_i^* is obtained by minimizing the following terms with respect to control u_i and $\bar{u}_i$,

$$\frac{\partial}{\partial u_i}\left[(\tau_i + e_i)\left(u_i - \bar{u}_i - \frac{r_i}{\tau_i + e_i}\alpha_0(x - \bar{x})\right)^2\right.$$
$$\left. + (\tau_i + \bar{\tau}_i + e_i + \bar{e}_i)\left(\bar{u}_i - \beta_0\frac{r_i + \bar{r}_i}{\tau_i + \bar{\tau}_i + e_i + \bar{e}_i}\bar{x}\right)^2\right] = 0,$$

which yields $u_i = \frac{r_i}{\tau_i + e_i}\alpha_0(x - \bar{x}) + \bar{u}_i$, where $\bar{u}_i = \beta_0(r_i + \bar{r}_i)/(\tau_i + \bar{\tau}_i + e_i + \bar{e}_i)\bar{x}$. Meanwhile, the mean field $\bar{x}(t)$ is derived by taking the expectation of the state dynamic equation in (6.16) and then solving the resulting differential equation for $\bar{x}(t)$.

Theorem 6.2 states that the optimal number of computation tasks ECN i must offload from the TA in a cooperative scenario is given by (6.24). This number minimizes the cost incurred by ECN i where α_0 and β_0 satisfy (6.25), and the mean field $\bar{x}(t)$ satisfies (6.26). It can be deduced that $u_i^*(t)$ not only depends on $x(t)$ but also on how much $x(t)$ exceeds $\bar{x}(t)$. Moreover, α_0 and β_0 capture how dependent $u_i^*(t)$ is on $x(t) - \bar{x}(t)$ and $\bar{x}(t)$, respectively.

6.4   Mean-Field-Type Game Based Computation Offloading Algorithms

This section presents the proposed algorithms that implement the MFTG-based computation offloading developed in the previous sections. A non-cooperative algorithm based on Theorem 6.1 is designed to simulate a scenario when the ECNs decide to minimize their own cost function. The algorithm can be implemented in a decentralized manner where each ECN decides for itself the optimal number of tasks to offload from the TA. Meanwhile, a cooperative algorithm based on Theorem 6.2 is designed for situations when the ECNs decide to minimize a global cost function. The algorithm can be implemented in a centralized manner where the TA decides for every ECN the optimal number of tasks to offload to the ECN. The proposed MFTG-based algorithms calculate the optimal solution $u_i^*(t)$ that corresponds to the portion of computation tasks that each ECN must offload in order to optimize its cost. As will be illustrated in Sect. 6.6, these algorithms improve the system cost and benefit-cost ratio of the local computing and dynamic greedy algorithms for computation offloading. Thus, the proposed MFTG-based algorithms can improve the targeted network performance.

Figure 6.2 illustrates the general procedure involved in the proposed algorithms. First, each ECN i determines its own cost coefficients r_i, τ_i, and e_i. Then, in the non-cooperative setting, ECN i computes the state and mean-state coupling coefficients α_i and β_i, while in the cooperative setting, the TA determines α_0 and β_0. These coefficients capture the effect of the state and mean-state to the optimal computation

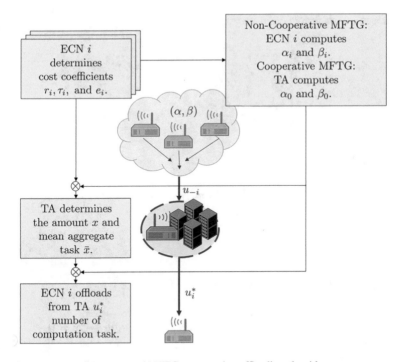

Fig. 6.2 Illustration of the proposed MFTG computation offloading algorithms

offloading control. At the same time, the TA determines the state $x(t)$ and mean-state $\bar{x}(t)$. Finally, the TA offloads a number of computation tasks to ECN i based on $u_i^*(t)$. The non-cooperative algorithm emulates a decentralized approach in which each ECN determines its own $u_i^*(t)$, while the cooperative algorithm follows a centralized approach in which the TA determines $u_i^*(t)$ of each ECN.

These algorithms require a sample period T_s and number of samples M, instead of a specified terminal time T. One main reason for this requirement is to avoid network parameter updates every time t, which can now be done every T_s. In addition, the cell dimension L and the number of ECNs N are required as well. The location of each node in the cell is limited within the area defined by $[0, L] \times [0, L]$. The TA is located at $z_0 = [L/2, L/2]$, while the location z_i of each ECN i is distributed randomly in the area. In addition, for each ECN i, its computing capability f_i and cost weights $w_{d,i}$ and $w_{e,i}$ are also defined.

6.4.1 Non-cooperative Computation Offloading

Since the non-cooperative solution using the direct approach stated in Theorem 6.1 assumes that each ECN has knowledge about the other ECNs, it has to be simplified in order to be implemented more practically. Let

$$\bar{\lambda} = \frac{1}{N} \sum_{j \in \mathcal{N}} \frac{r_j^2}{\tau_j + e_j} \alpha_j = \frac{1}{N} \sum_{j \in \mathcal{N}} \lambda_j,$$

$$\bar{\mu} = \frac{1}{N} \sum_{j \in \mathcal{N}} \frac{(r_j + \bar{r}_j)^2}{\tau_j + \bar{\tau}_j + e_j + \bar{e}_j} \beta_j = \frac{1}{N} \sum_{j \in \mathcal{N}} \mu_j.$$

Then, it follows that

$$\sum_{j \in \mathcal{N} \setminus i} \lambda_j = N\bar{\lambda} - \lambda_i,$$

$$\sum_{j \in \mathcal{N} \setminus i} \mu_j = N\bar{\mu} - \mu_i. \qquad (6.27)$$

Consequently, (6.21) can be rewritten as

$$\frac{r_i^2}{\tau_i + e_i} \alpha_i^2 + 2(N\bar{\lambda} - \lambda_i - r_0)\alpha_i - \rho_i = 0,$$

$$\frac{(r_i + \bar{r}_i)^2}{\tau_i + \bar{\tau}_i + e_i + \bar{e}_i} \beta_i^2 + 2(N\bar{\mu} - \mu_i - (r_0 + \bar{r}_0))\beta_i - (\rho_i + \bar{\rho}_i) = 0. \qquad (6.28)$$

Meanwhile, the mean values $\bar{r}_i$, $\bar{\tau}_i$, $\bar{e}_i$, and $\bar{\rho}_i$ can be found using the law of large numbers. It states that a sample average

$$\bar{S}_m = \frac{1}{m}(y_1 + \cdots + y_m),$$

converges to the expected value $\bar{y} = \mathbb{E}[y]$ as $m \to \infty$. Hence, the relationship between the parameters r_i, τ_i, e_i, and ρ_i and their respective expected values is given by

$$\lim_{m \to \infty} \frac{1}{m}(r_{i,1} + \cdots + r_{i,m}) = \bar{r}_i,$$

$$\lim_{m \to \infty} \frac{1}{m}(\tau_{i,1} + \cdots + \tau_{i,m}) = \bar{\tau}_i,$$

$$\lim_{m \to \infty} \frac{1}{m}(e_{i,1} + \cdots + e_{i,m}) = \bar{e}_i,$$

$$\lim_{m \to \infty} \frac{1}{m}(\rho_{i,1} + \cdots + \rho_{i,m}) = \bar{\rho}_i, \qquad (6.29)$$

$\forall i \in \mathcal{N}$.

Algorithm 2: Non-cooperative MFTG computation offloading

1: Set $M, T_s, L, N, z_i, f_i, w_{e,i}$, and $w_{d,i}$, $\forall i \in \mathcal{N}$.
2: Initialize $\bar{e}_i^{(0)}, \bar{\tau}_i^{(0)}, \bar{r}_i^{(0)}, \bar{\rho}_i^{(0)}, \bar{\lambda}$, and $\bar{\mu}$.
3: **for** $m = 1$ to M **do**
4:     **for** each ECN i in $\mathcal{N}$ **do**
5:         Compute e_i, τ_i, r_i using (6.1), (6.2), and (6.6), respectively. Compute α_i, β_i using (6.28).
6:         **for** each t in $0 \leq t \leq T_s$ **do**
7:             Observe and measure $x(t)$. Calculate $\bar{x}(t)$ using (6.22). Calculate $u_i^*(t)$ using (6.20).
8:         **end for**
9:         Update
$$\bar{e}_i^{(m)} = \tfrac{1}{m}(e_i + (m-1)\bar{e}_i^{(m-1)}),$$
$$\bar{\tau}_i^{(m)} = \tfrac{1}{m}(\tau_i + (m-1)\bar{\tau}_i^{(m-1)}),$$
$$\bar{r}_i^{(m)} = \tfrac{1}{m}(r_i + (m-1)\bar{r}_i^{(m-1)}),$$
$$\bar{\rho}_i^{(m)} = \tfrac{1}{m}(\rho_i + (m-1)\bar{\rho}_i^{(m-1)}).$$
10:         Update $\bar{\lambda}$ and $\bar{\mu}$ using (6.27).
11:     **end for**
12: **end for**

As a result, Algorithm 2 shows the non-cooperative computation offloading algorithm based on Theorem 6.1. After setting up some network parameters, each ECN i needs to initialize $\bar{r}_i, \bar{\tau}_i, \bar{e}_i, \bar{\lambda}$, and $\bar{\mu}$. Then, each ECN i determines r_i, τ_i, and e_i. Also, each ECN i estimates α_i and β_i using (6.21). Meanwhile, the TA broadcasts $x(t)$ and $\bar{x}(t)$ to the ECNs. Consequently, each ECN i can now calculate and offload from the TA the optimal offloading portion $u_i^*(t)$ that minimizes their own cost. Lastly, ECN i updates $\bar{r}_i, \bar{\tau}_i, \bar{e}_i, \bar{\lambda}$, and $\bar{\mu}$.

6.4.2 Cooperative Computation Offloading

Algorithm 3 implements the cooperative computation offloading based on Theorem 6.2. It starts with setting up some network parameters. Then, each ECN i initializes parameters such as $\bar{r}_i, \bar{\tau}_i$, and $\bar{e}_i$ and transmits them to the TA. Next, the TA computes α_0 and β_0 based on (6.25). Afterwards, the TA can now compute the optimal offloading control $u_i^*(t)$ of each ECN based on the values of $x(t)$ and $\bar{x}(t)$. Then, the TA offloads the corresponding number of computation tasks to each ECN i. Finally, the TA updates $\bar{r}_i, \bar{\tau}_i$, and $\bar{e}_i$.

Algorithm 3: Cooperative MFTG computation offloading

1: Set M, T_s, L, N, z_i, f_i, $w_{e,i}$, and $w_{d,i}$, $\forall i \in \mathcal{N}$.
2: Initialize $\bar{e}_i^{(0)}$, $\bar{\tau}_i^{(0)}$, and $\bar{r}_i^{(0)}$.
3: **for** $m = 1$ to M **do**
4:     **for** each ECN i in $\mathcal{N}$ **do**
5:         Compute e_i, τ_i, and r_i using (6.1), (6.2), and (6.6), respectively. Compute α_0 and β_0 using (6.25).
6:         **for** each t in $0 \leq t \leq T_s$ **do**
7:             Observe and measure $x(t)$. Calculate $\bar{x}(t)$ using (6.26). Calculate $u_i^*(t)$ using (6.24).
8:         **end for**
9:         Update
$$\bar{e}_i^{(m)} = \tfrac{1}{m}(e_i + (m-1)\bar{e}_i^{(m-1)}),$$
$$\bar{\tau}_i^{(m)} = \tfrac{1}{m}(\tau_i + (m-1)\bar{\tau}_i^{(m-1)}),$$
$$\bar{r}_i^{(m)} = \tfrac{1}{m}(r_i + (m-1)\bar{r}_i^{(m-1)}).$$
10:     **end for**
11: **end for**

6.5 Performance Evaluation

6.5.1 Baseline Approaches

To be able to evaluate the performance of the proposed MFTG computation offloading algorithms, they are compared with two typical algorithms in computation offloading. The first algorithm is the local computing based on [15]. It finds the number of computation tasks $x_0(t)$ that can be executed locally in the TA such that it satisfies the required deadline d_0, $x_0(t)/f_0 < d_0$. The cost function of the TA for local computing is defined by

$$J_{lo} = \mathbb{E}\left[\int_0^T (w_{d,0}\tau_{lo}x_0(t) + w_{e,0}e_{lo}x_0(t))\,dt \right], \qquad (6.30)$$

where $\tau_{lo} = 1/f_0$ refers to the number of time to execute a unit of computation task, $e_{lo} = \kappa_e f_0^2$ refers to the energy consumption per unit of computation task, and the constants $w_{d,0}$ and $w_{e,0}$ refer to the weights given by the TA to energy- and time-efficient optimization, respectively.

Another baseline algorithm used in this work is the dynamic greedy algorithm based on [15]. This algorithm finds the number of computation task $x_i(t)$ to be offloaded to ECN i that satisfies $x_i(t)/f_i < d_i$ where d_i is the deadline associated with $x_i(t)$. The cost function of ECN i for dynamic greedy computing is defined by

$$J_{dg,i} = \mathbb{E}\left[\int_0^T (w_{d,i}\tau_{dg,i}x_i(t) + w_{e,i}e_{dg,i}x_i(t))\,dt \right], \qquad (6.31)$$

where $\tau_{dg,i} = 1/f_i$ refers to the number of time to execute a unit of computation task, $e_{dg,i} = \kappa_e f_i^2$ refers to the energy consumption per unit of computation task, and the constants $w_{d,i}$ and $w_{e,i}$ refer to the weights given by ECN i to energy- and time-efficient optimization, respectively.

To bridge the gap between the baseline algorithms with linear cost functions and the proposed MFTG-based algorithms with quadratic cost functions, a quadratic term is added to the linear costs so that

$$J_{lo} = \mathbb{E}\left[\int_0^T (a_1 \xi_0 x_0(t) + a_2 \xi_0^2 x_0^2(t)) \, dt\right],$$
$$J_{dg,i} = \mathbb{E}\left[\int_0^T (a_1 \xi_i x_i(t) + a_2 \xi_i^2 x_i^2(t)) \, dt\right],$$

(6.32)

where $\xi_0 = w_{d,0}\tau_{lo} + w_{e,0}e_{lo}$, $\xi_i = w_{d,i}\tau_{dg,i} + w_{e,i}e_{dg,i}$, a_1 and a_2 as constants with $a_2 \ll a_1$.

6.5.2   Performance Metrics

The following metrics are calculated in order to compare the performance of the computation offloading approaches presented in this paper.

An offloading control fraction $p_i(t)$ is the ratio between the offloading control $u_i(t)$ and the state $x(t)$ of the TA, $p_i(t) = u_i(t)/x(t)$. Consequently, an optimal offloading control fraction $p_i^*(t)$ is written mathematically as

$$p_i^*(t) = \frac{u_i^*(t)}{x(t)},$$

(6.33)

where $u_i^*(t)$ is the optimal offloading control of ECN i.

The two main parameters in the MFTG formulation of computation offloading that limit the control of an ECN are energy consumption and computation or execution time. Consequently, the performance of the computation offloading methods are evaluated through energy efficiency and time efficiency. In this work, efficiency refers to how much computation task are executed per unit of network resource. Hence, energy efficiency is defined as the ratio between the number of computation tasks and the associated energy consumption. For an MECN with N ECNs, the network energy efficiency is written as

$$\eta_e = \frac{x(t)}{\sum_{i=1}^N \kappa_{e,i} f_i^2 u_i(t)},$$

(6.34)

where the ratio is taken between the total number of tasks at the TA and the total energy consumed by all the ECNs.

Meanwhile, time efficiency refers to the ratio between the number of computation tasks and the corresponding computation or execution time spent. For an MECN with N ECNs, the network time efficiency is given by

$$\eta_d = \frac{x(t)}{\sum_{i=1}^{N} \frac{u_i(t)}{f_i}}, \tag{6.35}$$

where the ratio is taken between the total number of tasks at the TA and the cumulative computation time of the tasks through the ECNs.

System cost is another way of comparing the computation offloading methods. It consists of the computation offloading cost and the overhead cost associated with each computation offloading algorithm. For both MFTG approaches, overhead exists between an ECN and the TA. Thus, the system costs for the non-cooperative and cooperative MFTG methods are given by

$$C_{nc} = \sum_{i=1}^{N} \left(\tilde{J}_i + 2\delta_i \theta_{i,0} \right),$$

$$C_{co} = \tilde{J}_0 + \sum_{i=1}^{N} 2\delta_i \theta_{i,0}, \tag{6.36}$$

where δ_i is the cost associated per overhead while $\theta_{i,0}$ is the number of overhead between ECN i and the TA. For the local computing algorithm, since the TA does not collaborate with any computing nodes, the overhead is zero. In the dynamic greedy offloading, overhead exists not only between the TA and ECNs but also between any two ECNs. Thus, the system costs for these two baseline approaches are

$$C_{lo} = J_{lo},$$

$$C_{dg} = \sum_{i=1}^{N} \left(J_{dg,i} + N\delta_{dg,i}\theta_{dg,i} \right), \tag{6.37}$$

where $\delta_{dg,i}$ is the cost associated per overhead while $\theta_{dg,i}$, is the number of overhead from ECN i to another computing node. In this work, overhead refers to the delay associated with the transmission time of overhead messages between any two computing nodes. For an overhead message of length b, the transmission time is b/r, where r is the rate at which the message is transmitted.

Lastly, to be able to compare the computational overhead and benefits of the proposed algorithms as well as the typical algorithms in computation offloading. The metric used to compare the algorithms is called the benefit-cost ratio B/C. The benefit B of each algorithm is the weighted sum of the energy and time efficiencies,

$$B = w_d \eta_d + w_e \eta_e, \tag{6.38}$$

where the constants w_d and w_e denote the weights given to the efficiencies and $w_d + w_e = 1$. The cost C used for each algorithm is the system cost defined previously.

6.6 Simulation Results and Discussion

6.6.1 Simulation Setup

The simulations presented in this work can be extended to networks containing multiple cells assuming that each cell operate independently of each other. That is, the TA of a cell can offload tasks only to ECNs located in its cell. Moreover, the interference between cells are minimized using techniques such as FDMA and SDMA. In addition, each simulation has been performed over 100 iterations, and the average of the results has been drawn in each figure.

Consider one network cell with an area of $150 \times 150 \, \text{m}^2$ containing one TA located at the center of the cell. The number of ECNs has been varied from 2 to 20. The location of each ECN is randomly distributed within the cell. Figure 6.3 shows the locations of the ECNs for the sparse MECN with $N = 5$ and the dense MECN with $N = 20$ utilized in the following simulations. The end users are located randomly within the cell. The number of end users are set to 50. The computation tasks arrive at the TA randomly, and the users are assumed to submit an average of 5 Tcycles of computation tasks.

Assume that the TA has a transmit power of 100 mW, a maximum incoming rate of computation task $R_0 = 10$ Gbps, and a maximum capacity C_0 of 10 Tb worth of computation tasks. Meanwhile, the computing nodes have a transmit power p_i of 100 mW and capacity C_i of 100 Gb worth of computation tasks. The computing capability f_i of each computing node is randomly selected from 10, 12, and 14 Tcycles/s. The cost weights $w_{d,i}$ and $w_{e,i}$ for the computation time and energy consumption are both set to 0.5. For SINR γ_i computations, the channel gain model used between any two nodes i and j is $g_{i,j} = d_{i,j}^{-\alpha}$ where $d_{i,j}$ denotes the distance between the two nodes and the path loss exponent $\alpha = 4$. Meanwhile, the background noise N_0 is set to -100 dBm. The quadratic cost constants for the baseline algorithms are $a_1 = 0.9$ and $a_2 = 0.1$.

6.6.2 Optimal Offloading Control

In the first of part the simulations, the optimal offloading control $u_i^*(t)$ of ECN i based on the feedback controls $u_{-i}^*(t)$ of other ECNs is computed using the MFTG computation offloading algorithms. Figure 6.4 shows the plots of $u_i^*(t)$ for the sparse

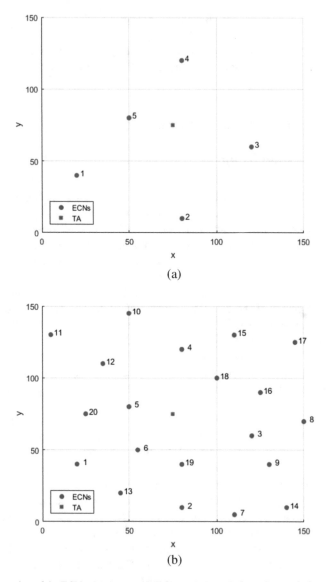

Fig. 6.3 Location of the ECNs. (a) A sparse MECN with $N = 5$. (b) A dense MECN with $N = 20$

MECN in both non-cooperative and cooperative MFTG scenarios as well as the number of computation tasks $x(t)$ at the TA. It can be noted from this figure that the two MFTG algorithms divide the computation tasks at the TA to the ECNs differently.

The partition of computation tasks among the ECNs is shown in Fig. 6.5 where each color denotes the particular share of an ECN. Figure 6.5a shows the average

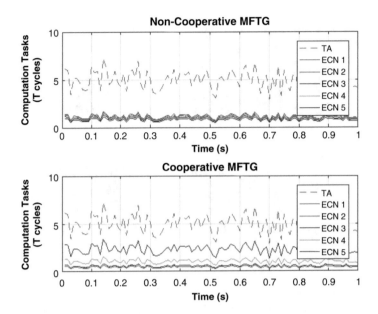

Fig. 6.4 Optimal offloading control of the ECNs in time domain

percentage of offloaded computation tasks from the TA to each ECN i in the sparse and dense MECN. In the sparse MECN, the non-cooperative MFTG approach distributes the tasks more evenly than the cooperative approach. Meanwhile, in the dense MECN, the distribution of tasks is more similar between the two MFTG offloading algorithms. The figure also implies that the offloading controls change accordingly when the number of ECNs is varied. Meanwhile, Fig. 6.5b presents the energy consumption per cycle of each ECN, and Fig. 6.5c shows the computation time per cycle contributed by each ECN.

Next, the effects of computing capability f_i and the cost weights $w_{d,i}$ and $w_{e,i}$ of ECN i to its optimal control $u_i^*(t)$ are investigated. While Fig. 6.6 shows $u_i^*(t)$ as a fraction of $x(t)$ averaged over time for ECN 1, the analyses that follow can be generalized to any ECNs. From the figure, it can be noticed that as the computing capability f_i of an ECN increases, the average percentage of aggregate computation task it offloads rises up to a certain point, then it decreases. The reason for this trend is the compromise between minimizing computation time and energy consumption. When f_i is low, the energy consumption of ECN i is also low; however, the computation time to execute the offloaded tasks is high. As f_i becomes higher, the energy consumption of an ECN increases while the computation time to execute the offloaded tasks becomes lesser.

Meanwhile, as computation time is given more weight by increasing its weight from 0.1 to 0.9, the curve shifts to the right. This means that as an ECN prioritizes minimizing computation time, the computing capability at which it can afford to offload the highest percentage of the aggregate computation task increases.

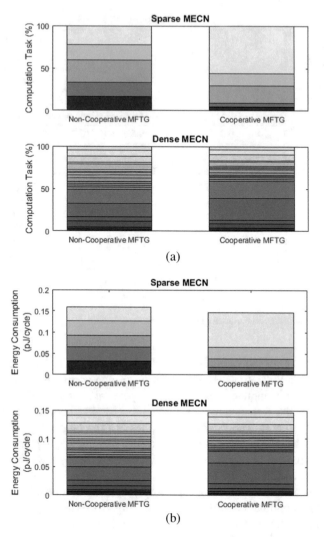

Fig. 6.5 Partition of aggregate computation task for each ECN. (**a**) Average percentage of computation task. (**b**) Average energy consumption of ECNs. (**c**) Average computation time of ECNs

However, as more weight is given to energy consumption from 0.1 to 0.9, the curve shifts to the left. That is, to lower the energy consumption of an ECN, the computing capability at which its offloading percentage is at the highest decreases.

In summary, an ECN with lower computing capability offloads more from the TA if minimizing the energy consumption is more critical, as shown by the red curves. However, if the priority is to minimize computation time, then an ECN with higher computing capability offloads more from the TA, as shown by the blue curves.

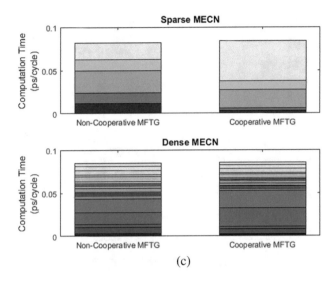

Fig. 6.5 (continued)

6.6.3 Network Efficiency

The energy efficiency η_e using different computation offloading approaches are compared in Fig. 6.7a. From the figure, the cooperative MFTG (CMFTG) approach has better η_e than the non-cooperative MFTG (NCMFTG) approach. However, both MFTG algorithms have higher η_e than the local and dynamic greedy algorithms. This is one of the reasons that justifies the significance of computation offloading in MECN.

Meanwhile, the time efficiency η_d of the network under different computation offloading approaches are displayed in Fig. 6.7b. It can be concluded from the figure that MFTG computation offloading approaches maintain a competitive η_d against the dynamic greedy algorithm.

Hence, the MFTG offloading algorithms can be as efficient as the dynamic greedy algorithm which requires full knowledge of the characteristics of all the ECNs. In the following subsection, the system costs of the computation offloading algorithms are compared.

6.6.4 System Cost and Benefit-Cost Ratio

Figure 6.8 presents the system cost sustained by each computation algorithm investigated in this work. The system cost of the dynamic greedy approach is higher than both the MFTG approaches because the overhead required to implement the greedy algorithm is larger than the overhead required by the MFTG approaches. The

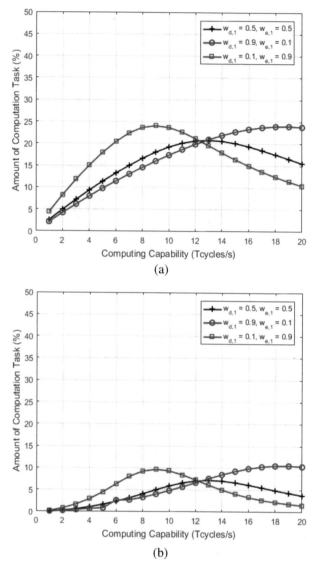

Fig. 6.6 Effect of computing capability and cost weights to the optimal computation offloading control. (**a**) Non-cooperative MFTG. (**b**) Cooperative MFTG.

system cost of the local computing approach is shown for comparison purposes even though it does not require the use of ECNs. Between the two MFTG approaches, the cooperative approach has lower system cost than the non-cooperative approach when the number of ECNs is lower. However, as the number of ECNs increases, the system cost of the non-cooperative approach becomes lower than that of the cooperative approach.

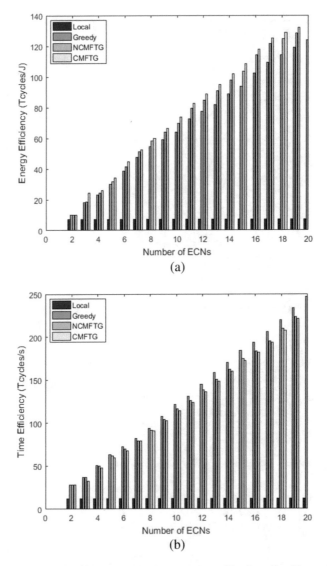

Fig. 6.7 Average network efficiency of the computation offloading algorithms. (**a**) Average network energy efficiency. (**b**) Average network time efficiency

Figure 6.9 shows the benefit-cost ratio B/C for each computation offloading approaches. It is evident that the non-cooperative MFTG approach has the best B/C, followed by the cooperative MFTG approach. The benefit of the MFTG approaches are contributed by the energy- and time-efficient partition of computation tasks as well as the low number of network overhead required to implement the offloading.

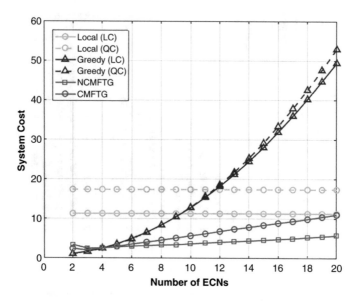

Fig. 6.8 The average network cost of the computation offloading algorithms

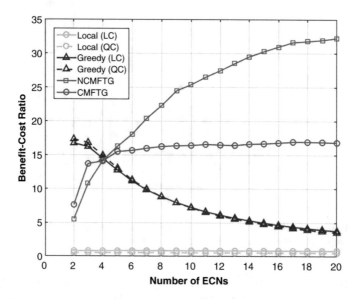

Fig. 6.9 Benefit-cost ratio of the computation offloading algorithms

Moreover, the system cost and benefit-cost ratio of the local computing and dynamic greedy algorithms with quadratic cost (QC) are almost equivalent to the system cost and benefit-cost ratio of the original linear cost (LC). Therefore, the form of the cost function does not affect the performance of the algorithms significantly since the main difference between the cost functions of the proposed work and the baseline algorithms is the overhead cost.

6.7 Related Works

Offloading of computation-intensive tasks from mobile devices to MECNs has garnered a lot of interests in the research community. In this section, these computation offloading methods are briefly described.

Various game theoretic methods have been applied to model computation offloading among many computing units. In [16], the authors utilized a game theoretic approach to computation offloading problem among mobile users in a multi-user, multi-channel wireless MECN. Meanwhile, in order to utilize the computation resources in the cloud, collaborative computation offloading between the centralized cloud server and the MEC servers was studied in [17].

Many research have jointly optimized computation offloading with other network technologies and issues. The authors of [18] formulated computation offloading among mobile devices as a joint optimization of the radio and computation resources that minimizes a user's energy consumption while satisfying latency requirements. An energy-efficient dynamic offloading and resource scheduling formulated as a minimization problem was investigated in [19]. In [20], interference management was integrated in computation offloading and formulated together as an optimization problem. To reduce execution delay, computation offloading was integrated with cache placement in MEC to store and share popular computation results to mobile users [21].

Several works have focused on integrating computation offloading feature in networks involving wireless power. Computation offloading in mobile cloud computing powered by wireless energy transfer was studied in [22]. The authors proposed the use of CPU-cycle statistics information and channel state information to enforce policies that maximize the probability of successful computation of data subject to the energy harvesting and latency constraints. In [23], the authors combined the concepts of MEC with wireless power transfer so that the MEC access point can transmit wireless power to mobile users which can be used for local computing. Then, the authors of [15] proposed a Lyapunov optimization-based dynamic algorithm for MEC with energy-harvesting devices that jointly decides on the offloading, CPU frequency, and transmit power.

Energy-efficient computation offloading algorithms have been the focus of several works as well. An energy-efficient computation offloading scheme was proposed in [24] where the energy consumption of the offloading system was minimized while still satisfying the latency requirements of the tasks. Meanwhile, energy-efficient task offloading in software defined ultra-dense network was investigated in [25].

Partial offloading where only a part of an application is offloaded to computing entities has been studied by several works. The authors of [26] considered partial offloading due to limited bandwidth in wireless networks. Also, the authors of [27] proposed a cooperative partial computation offloading between cloud computing and MEC-enabled IoT.

Before concluding this section, some works involving MFTGs are worth mentioning. In [28], energy storage problem in a microgrid was formulated as an MFTG. The mean and variance of the energy level were added to the cost function and used MFTG to keep track and maintain the desired energy level in the microgrid. Meanwhile, MFTG was utilized as a particle filter for video-based vehicular tracking in Intelligent Traffic Systems (ITS) [29]. A mean field term was included in the formulation to provide accurate and robust state (i.e., vehicle position) prediction. In [30], MFTG was applied in blockchain token economics. This work introduced variance in the utility function to capture the risk of cryptographic tokens associated with the uncertainties of technology adoption, network security, regulatory legislation, and market volatility.

The main difference of the proposed MFTG-based work is that computation offloading in MECN has been formulated as an MFTG in which each computing node has a desired level of computation tasks it can handle. This level is dictated by the energy consumption and computing capability of the computing node. Moreover, this work utilizes a direct approach that does not require solving coupled partial differential equations to solve for the optimal computation offloading strategy of each computing node. Lastly, non-cooperative and cooperative scenarios among the computing nodes are both considered and investigated.

6.8　Conclusion

Multi-access edge computing networks (MECN) reduce the latency inherent in cloud computing networks by performing the tasks in an edge network near the network users rather than in a cloud network. Computation offloading is one of the services in an MECN in which computation-intensive tasks in a computing node may be offloaded to other computing nodes in the network. In this work, computation offloading problem has been formulated using mean-field-type game (MFTG). Then, non-cooperative and cooperative computation offloading algorithms have been proposed. These algorithms search for the optimal computation offloading control (i.e., optimal number of computation tasks) of each computing node in an MECN. The non-cooperative algorithm is a decentralized approach since each computing node determines its own offloading control. Nevertheless, the cooperative algorithm is a centralized approach in which the network determines the offloading control of each computing node. Lastly, the simulation results have indicated that MFTG is an effective way to model computation offloading in MECNs.

References

1. X. Chen, Decentralized computation offloading game for mobile cloud computing. IEEE Trans. Distrib. Parallel Comput. **26**(4), 974–983 (2015)
2. T. Soyata, R. Muraleedharan, C. Funai, M. Kwon, W. Heinzelman, Cloud-vision: real-time face recognition using a mobile-cloudlet-cloud acceleration architecture, in *Proc. IEEE Symposium on Computers and Communications*, July 2012, pp. 59–66
3. M. Greer, *Electricity Cost Modeling* (Elsevier, Oxford, 2010)
4. W. Zhang, Y. Wen, K. Guan, D. Kilper, H. Luo, D. Wu, Energy-optimal mobile cloud computing under stochastic wireless channel. IEEE Trans. Wirel. Commun. **12**(9), 4569–4581 (2013)
5. S. Samarakoon, M. Bennis, W. Saad, M. Debbah, M. Latva-Aho, Ultra dense small cell networks: turning density into energy efficiency. IEEE J. Sel. Areas Commun. **34**(5), 1267–1280 (2016)
6. J.M. Lasry, P.L. Lions, Mean field games. Jpn. J. Math. **2**(1), 229–260 (2007)
7. B. Jovanovic, Selection and the evolution of industry. Econometrica **50**(3), 649–670 (1982)
8. B. Jovanovic, R.W. Rosenthal, Anonymous sequential games. J. Math. Econ. **17**(1), 77–87 (1988)
9. J. Bergin, D. Bernhardt, Anonymous sequential games with aggregate uncertainty. J. Math. Econ. **21**(6), 543–562 (1992)
10. B. Djehiche, A. Tcheukam, H. Tembine, Mean-field-type games in engineering. AIMS Electron. Electr. Eng. **1**(1), 18–73 (2017)
11. A. Tcheukam, H. Tembine, Mean-field-type games for distributed power networks in presence of prosumers, in *Proc. Chinese Control and Decision Conference (CCDC), Yinchuan*, May 2016, pp. 446–451
12. A. Siwe, H. Tembine, Network security as public good: a mean-field-type game theory approach, in *Proc. 13th International Multi-Conference on Systems, Signals & Devices (SSD), Leipzig*, Mar 2016, pp. 601–606
13. B. Djehiche, A. Tcheukam, H. Tembine, A mean-field game of evacuation in multilevel building. IEEE Trans. Autom. Control **62**(10), 5154–5169 (2017)
14. T.E. Duncan, H. Tembine, Linear-quadratic mean-field-type games: a direct approach. Games **9**(1), 7–24 (2018)
15. Y. Mao, J. Zhang, K. Letaief, Dynamic computation offloading for mobile-edge computing with energy harvesting devices. IEEE J. Sel. Areas Commun. **34**(12), 3590–3605 (2016)
16. X. Chen, L. Jiao, W. Li, X. Fu, Efficient multi-user computation offloading for mobile-edge cloud computing. IEEE/ACM Trans. Netw. **24**(5), 2795–2808 (2016)
17. H. Guo, J. Liu, Collaborative computation offloading for multiaccess edge computing over fiber-wireless networks. IEEE Trans. Veh. Technol. **67**(5), 4514–4526 (2018)
18. S. Sardellitti, G. Scutari, S. Barbarossa, Joint optimization of radio and computational resources for multicell mobile-edge computing. IEEE Trans. Signal Inf. Process. Netw. **1**(2), 89–103 (2015)
19. S. Guo, B. Xiao, Y. Yang, Y. Yang, Energy-efficient dynamic offloading and resource scheduling in mobile cloud computing, in *Proc. IEEE INFOCOM, San Francisco, CA*, April 2016
20. C. Wang, F. Yu, C. Liang, Q. Chen, L. Tang, Joint computation offloading and interference management in wireless cellular networks with mobile edge computing. IEEE Trans. Veh. Technol. **66**(8), 7432–7445 (2017)

21. S. Yu, R. Langar, X. Fu, L. Wang, Z. Han, Computation offloading with data caching enhancement for mobile edge computing. IEEE Trans. Veh. Technol. **67**(11), 11098–11112 (2018)
22. C. You, K. Huang, H. Chae, Energy efficient mobile cloud computing powered by wireless energy transfer. IEEE J. Sel. Areas Commun. **34**(5), 1757–1771 (2016)
23. F. Wang, J. Xu, X. Wang, S. Cui, Joint offloading and computing optimization in wireless powered mobile-edge computing systems. IEEE Trans. Wirel. Commun. **17**(3), 1784–1797 (2018)
24. K. Zhang, Y. Mao, S. Leng, Q. Zhao, L. Li, X. Peng, L. Pan, S. Maharjan, Y. Zhang, Energy-efficient offloading for mobile edge computing in 5G heterogeneous networks. IEEE Access **4**, 5896–5907 (2016)
25. M. Chen, Y. Hao, Task offloading for mobile edge computing in software defined ultra-dense network. IEEE J. Sel. Areas Commun. **36**(3), 587–597 (2018)
26. Y. Wang, M. Sheng, X. Wang, L. Wang, J. Li, Mobile-edge computing: partial computation offloading using dynamic voltage scaling. IEEE Trans. Commun. **64**(10), 4268–4282 (2016)
27. Z. Ning, P. Dong, X. Kong, F. Xia, A cooperative partial computation offloading scheme for mobile edge computing enabled internet of things. IEEE Internet Things J. (2018). https://doi.org/10.1109/JIOT.2018.2868616
28. J. Barreiro-Gomez, T.E. Duncan, H. Tembine, Linear-quadratic mean-field-type games-based stochastic model predictive control: a microgrid energy storage application, in *Proc. 2019 American Control Conference (ACC), Philadelphia, PA,* July 2019
29. J. Gao, H. Tembine, Distributed mean-field-type filter for vehicle tracking, in *Proc. 2017 American Control Conference (ACC), Seattle, WA,* May 2017
30. J. Barreiro-Gomez, H. Tembine, Blockchain token economics: a mean-field-type game perspective. IEEE Access **7**, 64603–64613 (2019)

Printed in the United States
by Baker & Taylor Publisher Services